Longman Top Pocket Series

KU-051-583

Pocket Life Sciences Dictionary

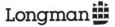

Foreword

With the aid of this little book, you can now unravel the mystery of those technical terms which scientists use to describe the living world around us. Simple explanations are given for those complicated words you hear, see, and read, and cover subjects ranging from psychology, medicine, and botany to soil science, palaeontology, and even statistics.

The definitions are clear and easy to understand, and, combined with several pages of attractive illustrations and tables, make this pocket-sized reference book a must for the family bookshelf, and also a useful aid for the student of any of the life sciences.

Cross references to illustrations are denoted by two symbols – an eye ● and a hand ☞. An eye refers to an illustration or table on the following page, and a hand, followed by a word in small capitals, refers to an illustration at the main entry for that word elsewhere in the dictionary.

abaxial *adj* situated outside or directed away from the axis of an organ, plant part, or organism — compare ADAXIAL

abdomen *n* 1 (the cavity of) the part of the body between the thorax and the pelvis that contains the liver, gut, etc 2 the rear part of the body behind the thorax in an insect or other arthropod

abducens nerve *n* either of the 6th pair of cranial nerves which are motor nerves supplying muscles of the eye

aberrant *adj* diverging from the usual or natural type

abiogenesis *n* the supposed spontaneous origination of living organisms directly from lifeless matter

abiotic *adj* not involving or produced by living organisms

abnormal psychology *n* the psychology of mental disorder

abomasum *n, pl* **abomasa** the fourth or true digestive stomach of a ruminant mammal

aboriginal *adj* indigenous

abort *vi* 1 to expel a premature nonviable foetus 2 to fail to develop completely; shrink away ~ *vt* 1 to induce the abortion of (a foetus) 2 to stop in the early stages <~ *a disease*>

abortifacient *n or adj* (a drug or other agent) inducing abortion

abortion *n* 1 the spontaneous or induced expulsion of a foetus 2 (the result of) an arresting of development of a part, process, etc

abortive *adj* imperfectly formed or developed

aboulia, abulia *n* pathological loss of willpower

abrasion *n* an area of the skin or mucous membrane that has been scraped, rubbed, or grazed

abreaction *n* the release of tension due to a repressed emotion by means of reliving the situation in which it orig occurred

abscess *n* a pocket of pus surrounded by inflamed tissue

abscise *vb* to separate by abscission

abscisic acid *n* a plant hormone that typically promotes leaf abscission and dormancy

abscisin *also* **abscissin** *n* abscisic acid or a similar plant hormone

abscission *n* the natural separation of flowers, leaves, etc from plants at a special separation layer

absorb *vt* 1 to take in and make part of an existent whole 2 to suck up or take up <*plant roots* ~ *water*>

acariasis *n* infestation with or disease caused by mites

acarid *n* a typical mite or other related arachnid

accommodation *n* the (range of) automatic adjustment of the eye, esp by changes in the amount by which the lens bends light, for seeing at different distances

acephalous *adj* lacking a head or having the head reduced

1

ace

acetabulum *n, pl* **acetabulums, acetabula 1a** the cup-shaped socket in the hipbone into which the head of the thighbone fits **b** the cavity in the body of an insect into which its leg fits **2** a round sucker of a leech or other invertebrate

acetylcholine *n* a neurotransmitter released esp at autonomic nerve endings

acetyl-coA *n* acetyl coenzyme A

acetyl coenzyme A *n* a compound formed as an essential intermediate in the metabolism of most living cells

achene *n* a small dry indehiscent 1-seeded fruit (eg that of the dandelion)

acid *n* a usu water-soluble substance that reacts with a base to form a salt and turns blue litmus red

acidophilic, acidophil *adj* **1** staining readily with acid dyes **2** preferring or thriving in an acid environment

acidosis *n* a disorder in which the blood, tissues, etc are unusually acid

acoelomate *adj, of an animal* having no coelom

acromegaly *n* abnormal enlargement of the hands, feet, and face caused by excessive production of growth hormone by the pituitary gland

acrophobia *n* abnormal dread of being at a great height

ACTH *n* ADRENOCORTICOTROPHIC HORMONE

actin *n* a protein found in muscle and other cells that combines with myosin in producing muscular contraction

actinomorphic *also* **actinomorphous** *adj, of an organism or part* radially symmetrical

actinomycete *n* any of an order of filamentous or rod-shaped bacteria

actinozoan *n* an anthozoan

action potential *n* a momentary change in the electrical potential across the membrane of a (nerve) cell resulting from activation by a stimulus

activate *vt* to make (more) active or reactive, esp in chemical or physical properties: eg to aerate (sewage) so as to favour the growth of organisms that decompose organic matter

active transport *n* movement of a chemical substance across a (cell) membrane in living tissue by the expenditure of energy

aculeate *adj* having a sting <*an ~ insect*>

acuminate *adj* tapering to a slender point ☞ PLANT

acute *adj* **1** *esp of an illness* having a sudden severe onset and short course — compare CHRONIC 1 **2** ending in a sharp point

adapt *vb* to make or become fit, often by modification

adaptation *n* **1** adjustment of a sense organ to the intensity or quality of stimulation **2** modification of (the parts of) an organism fitting it better for existence and successful breeding

adaxial *adj* situated on the same side as or facing the axis of an organ, plant part, or organism — compare ABAXIAL

addict[1] *vt* to cause (an animal or human) to become physiologically dependent upon a habit-forming drug

addict[2] *n* one who is addicted to a drug

adenine *n* a purine base that is 1 of the 4 bases whose order in a DNA or RNA chain codes genetic information — compare CYTOSINE, GUANINE, THYMINE, URACIL

adenoid *adj or n* (of) an enlarged mass of lymphoid tissue at the back of the pharynx, often obstructing breathing — usu pl with sing. meaning

adenoma *n, pl* **adenomas, adenomata** a benign tumour of a glandular structure or of glandular origin

adenosine *n* a nucleoside containing adenine

adenosine diphosphate *n* ADP

adenosine monophosphate *n* 1 AMP **2** CYCLIC AMP

adenosine triphosphate *n* ATP

adipose *adj* of animal fat; fatty

adipose tissue *n* connective tissue in which fat is stored

adjuvant *n* sthg that helps or makes esp medical treatment more effective

ADP *n* a derivative of adenine that is reversibly converted to ATP for the storing of cellular energy [adenosine *di*phosphate]

adrenal *adj* **1** adjacent to the kidneys **2** of or derived from adrenal glands

adrenal gland *n* an endocrine gland near the front of each kidney with a cortex that secretes steroid hormones and a medulla that secretes adrenalin

adrenalin, adrenaline *n* a hormone produced by the adrenal gland that occurs as a neurotransmitter in the sympathetic nervous system and that stimulates the heart and causes constriction of blood vessels and relaxation of smooth muscle

adrenocorticotrophic hormone *n* a hormone of the front lobe of the pituitary gland that stimulates the adrenal cortex

adsorption *n* the adhesion in an extremely thin layer of molecules of gases, liquids, etc to the surfaces of solids or liquids

adventitious *adj* occurring sporadically or in an unusual place <~ *buds on a plant*>

aerial *adj* growing in the air rather than in the ground or water <~ *roots*>

aerobe *n* an organism (eg a bacterium) that lives only in the presence of oxygen

aestivate, *NAm also* **estivate** *vi, of animals, esp insects* to pass the summer in a state of torpor — compare HIBERNATE

aestivation *n* the arrangement of floral parts in a bud — compare VERNATION

aetiology, *chiefly NAm* **etiology** *n* (the study of) the causes or origin, specif of a disease or abnormal condition

afferent *adj* bearing or conducting (nervous impulses) inwards (towards the brain) — compare EFFERENT

aflatoxin *n* any of several poisons that are produced by moulds (eg in badly stored peanuts) and cause (liver) cancers

afterbirth *n* the placenta and foetal membranes expelled after delivery of a baby, young animal, etc

afterimage *n* a usu visual sensation remaining after external stimulation (eg of the retina) has ceased

aftertaste *n* persistence of a flavour or impression after the stimulus is no longer present

agamic *adj* asexual, parthenogenetic

agar-agar *n* a gelatinous extract from any of various red algae used esp in culture media or as a gelling agent in food

agaric *n* any of a family of fungi (eg the common edible mushroom) with an umbrella-shaped cap

age[1] *n* **1** a period in (eg human or geological) history <the ~ of reptiles> **2** an individual's develop-

ment in terms of the years required by an average individual for similar development <a mental ~ of 6>

age[2] *vi* to become mellow or mature; ripen

agent *n* a chemically, physically, or biologically active substance

agglomerate *adj* gathered into a ball, mass, or cluster

agglutination *n* the collection of red blood cells or other minute suspended particles in clumps, esp as a response to a specific antibody

agglutinin *n* a substance producing biological agglutination

aggregate[1] *adj* **1** *of a flower* clustered in a dense mass or head **2** *of a fruit* formed from the several ovaries of a single flower

aggregate[2] *n* a clustered mass of individual particles of various shapes and sizes that is considered to be the basic structural unit of soil

aggression *n* **1** a forceful action esp when intended to dominate **2** hostile, injurious, or destructive behaviour or outlook

agonist *n* **1** a muscle that is restricted by the action of an antagonistic muscle with which it is paired **2** a substance capable of combining with a receptor on the surface of a (nerve) cell and initiating a reaction

agoraphobia *n* abnormal dread of being in open spaces

agranulocyte *n* any of various

4

white blood cells with cytoplasm that does not contain conspicuous granules — compare GRANULOCYTE

A-horizon *n* mineral material mixed with humus forming the surface layer of soil

alanine *n* an amino acid found in most proteins

alar *adj* of or like a wing

albino *n, pl* **albinos** an organism with (congenitally) deficient pigmentation; *esp* a human being or other animal with a (congenital) lack of pigment resulting in a white or translucent skin, white or colourless hair, and eyes with a pink pupil

albumin, albumen *n* any of numerous proteins that occur in large quantities in blood plasma, milk, egg white, plant fluids, etc and are coagulated by heat

albuminous *adj* relating to, containing, or like albumen or albumin

albuminuria *n* the (abnormal) presence of albumin in the urine, usu symptomatic of kidney disease

alcohol *n* a colourless volatile inflammable liquid that is the intoxicating agent in fermented and distilled drinks and is used also as a solvent

aldosterone *n* a steroid hormone produced by the adrenal cortex that affects the salt and water balance of the body

aleurone *n* minute granules of protein in (the endosperm of) seeds

alexia *n* (partial) loss of the ability to read, owing to brain damage — compare APHASIA, DYSLEXIA

alga *n, pl* **algae** *also* **algas** any of a group of chiefly aquatic nonvascular plants (eg seaweeds and pond scums); *also* BLUE-GREEN ALGA ☞ EVOLUTION, PLANT

alginic acid *n* an insoluble colloidal acid found in the cell walls of brown algae

alimentary canal *n* the tubular passage that extends from the mouth to the anus and functions in the digestion and absorption of food

alkalosis *n* a medical disorder in which the blood, tissues, etc are abnormally alkaline

allantois *n, pl* **allantoides** a vascular foetal membrane that in placental mammals is closely attached to the chorion in the formation of the placenta

allele *n* any of (the alternative hereditary characters determined by) 2 or more genes that occur as alternatives at a given place on a chromosome

allelomorph *n* an allele

allergen *n* a substance that induces allergy

allergic *adj* of or inducing allergy

allergy *n* **1** altered bodily reactivity to an antigen in response to a first exposure <*his bee-venom ~ may make a second sting fatal*> **2** exag-

5

gerated reaction by sneezing, itching, skin rashes, etc to substances that have no such effect on the average individual

allochthonous, allocthonous *adj, of a plant, animal, or substance* entering a particular ecological region from an outside source

allogamous *adj* reproducing by cross-fertilization

allograft *n* a graft between 2 genetically unlike members of the same species

alopecia *n* usu abnormal baldness in humans or loss of wool, feathers, etc in animals

alpha-receptor *n* a receptor for neurotransmitters (eg adrenalin) in the sympathetic nervous system whose stimulation is associated esp with the constriction of small blood vessels and an increase in blood pressure — compare BETA-RECEPTOR

alpha wave *n* a variation in the electroencephalographic record of the electrical activity of the brain of a frequency of about 10Hz that is often associated with states of waking relaxation

Alpine *adj, often not cap* of or growing in the Alps or other mountains esp above the tree line

alternate *adj, of plant parts* arranged singly first on one side and then on the other of an axis — compare OPPOSITE ☞ PLANT

alternation of generations *n* the occurrence of 2 or more usu alternating sexual and asexual forms produced during the life cycle of a plant or animal

altricial *adj, of a bird* (having young) needing care for some time after birth — compare PRECOCIAL

alula *n, pl* **alulae** BASTARD WING

alveolar *adj* of, resembling, made up of, or having alveoli or an alveolus

alveolus *n, pl* **alveoli** a small cavity or pit: eg **a** a socket for a tooth **b** an air cell of the lungs **c** a cell or compartment of a honeycomb

amaurosis *n, pl* **amauroses** decay of sight, esp due to neurological disease, without obvious change or damage to the eye

ambivert *n* a person with both extroverted and introverted characteristics

amblyopia *n* poor sight without obvious change or damage to the eye, esp due to toxic effects or dietary deficiencies

amenorrhoea, *chiefly NAm* **amenorrhea** *n* abnormal absence of the menstrual discharge

amentia *n* (congenital) mental deficiency

amino acid *n* any of various organic acids containing an amino group and occurring esp in linear chains as the chief components of proteins

amitosis *n* cell division by simple cleavage of the nucleus and division of the cytoplasm without spindle formation or appearance of chromosomes

ammonification *n* decomposition, esp of nitrogenous organic matter by bacteria, with production of ammonia or ammonium compounds

ammonite *n* a flat spiral fossil shell of a mollusc abundant esp in the Mesozoic age

amnesia *n* a (pathological) loss of memory

amniocentesis *n* the insertion of a hollow needle into the uterus of a pregnant female, esp to obtain amniotic fluid (eg for the detection of chromosomal abnormality)

amnion *n, pl* **amnions, amnia** a thin membrane forming a closed sac containing the watery fluid in which an embryo is immersed

amoeba, *chiefly NAm* **ameba** *n, pl* **amoebas, amoebae** any of various protozoans with lobed pseudopodia and without permanent organelles that are widely distributed in water and wet places

amoeboid, *chiefly NAm* **ameboid** *adj* (moving by means of protoplasmic flow) like an amoeba

AMP *n* a mononucleotide of adenine that is reversibly converted in cells to ADP and ATP; *also* CYCLIC AMP [adenosine *monophosphate]

amphibian *n, pl* **amphibians,** *esp collectively* **amphibia** an amphibious organism; *esp* a frog, toad, newt, or other member of a class of cold-blooded vertebrates intermediate in many characteristics between fishes and reptiles ⟶ EVOLUTION

amphibious *adj* able to live both on land and in water

amphimictic *adj* capable of (producing fertile offspring by) interbreeding

amphimixis *n, pl* **amphimixes** (the union of germ cells in) sexual reproduction — compare APOMIXIS

amplexus *n* the mating embrace of a frog or toad during which eggs are shed into the water and there fertilized

ampulla *n, pl* **ampullae** a saclike anatomical swelling or pouch

amylase *n* an enzyme that accelerates the hydrolytic breakdown of starch and glycogen

amyloid *n* a firm waxy substance deposited in animal organs under abnormal conditions

amylopsin *n* the amylase of the pancreatic juice

amylose *n* (a component or hydrolytic product of) starch or a similar polysaccharide

anabiosis *n, pl* **anabioses** a state of suspended animation induced in some organisms by desiccation

anabolic steroid *n* any of several synthetic steroid hormones that cause a rapid increase in the size and weight of skeletal muscle

anabolism *n* constructive metabolism involving the use of energy by

7

a living organism to make proteins, fats, etc from simpler materials — compare CATABOLISM

anaemia, *chiefly NAm* **anemia** *n* a condition in which the blood is deficient in red blood cells, haemoglobin, or total volume

anaerobe *n* an organism (eg a bacterium) that lives only in the absence of oxygen

anaesthesia, *chiefly NAm* **anesthesia** *n* loss of sensation, esp loss of sensation of pain, resulting either from injury or a disorder of the nerves or from the action of drugs

anaesthetic, *chiefly NAm* **anesthetic** *n* a substance that produces anaesthesia, eg so that surgery can be carried out painlessly

anal *adj* **1** of or situated near the anus **2** of or characterized by (parsimony, meticulousness, or other personality traits typical of) the stage of sexual development during which the child is concerned esp with its faeces — compare ORAL, GENITAL

analgesia *n* insensibility to pain without loss of consciousness

analyse, *NAm chiefly* **analyze** *vt* **1** to study the nature and relationship of the parts of by analysis **2** to psychoanalyse

analysis *n, pl* **analyses 1** separation of a whole (eg an organism, organ, environment, or chemical) into its component parts **2** psychoanalysis

anaphase *n* the stage of mitosis and meiosis in which the chromosomes move towards the poles of the spindle

anaphylaxis *n, pl* **anaphylaxes** a sometimes fatal reaction to drugs, insect venom, etc due to hypersensitivity resulting from earlier contact

anaplasia *n* reversion of cells to a more primitive or undifferentiated form

anatomy *n* **1** (a treatise on) the morphology of the structure of organisms **2** dissection **3** structural make-up, esp of (a part of) an organism 🔊

ancestor *n* a progenitor of a more recent (species of) organism

androecium *n, pl* **androecia** all the stamens collectively in the flower of a seed plant

androgen *n* a male sex hormone (eg testosterone)

androgynous *adj* having characteristics of both the male and female forms

anemophilous *adj* (usually) wind-pollinated

aneurysm *also* **aneurism** *n* a permanent blood-filled swelling of a (large) diseased blood vessel (eg the aorta)

angina pectoris *n* brief attacks of intense chest pain, esp on exertion, precipitated by deficient oxygenation of the heart muscles

angiosperm *n* any of a class of

vascular plants having the seeds enclosed by the ovary (eg buttercups, orchids, roses, oaks, or grasses) — compare GYMNOSPERM ⌐ PLANT

angiotensin *n* either of 2 related hormones that influence the fluid balance of the body — compare RENIN

anima *n* an individual's true inner self reflecting archetypal ideals of conduct; *also* an inner feminine part of the male personality — used in Jungian psychology; compare ANIMUS, PERSONA

animal *n* **1** any of a kingdom of living things typically differing from plants in their capacity for spontaneous movement, esp in response to stimulation **2** any of the lower animals as distinguished from human beings

animalcule *n* a minute usu microscopic organism

animal kingdom *n* that one of the 3 basic groups of natural objects that includes all living and extinct animals — compare MINERAL KINGDOM, PLANT KINGDOM

animus *n* an inner masculine part of the female personality — used in Jungian psychology; compare ANIMA

anion *n* the ion in an electrolyzed solution that migrates to the anode; *broadly* a negatively charged ion

anisogamous *adj* having or being

produced by the fusion of gametes which differ usu in size

anisotropic *adj* assuming different positions in response to external stimuli

anlage *n*, the foundation of a subsequent development; *specif* a primordium

annelid *n* any of a phylum of usu elongated segmented invertebrates (eg earthworms and leeches)

annual[1] *adj*, of a plant completing the life cycle in 1 growing season

annual[2] *n* sthg lasting 1 year or season; *specif* an annual plant

annual ring *n* the layer of wood produced by a single year's growth of a woody plant

annulus *n*, *pl* **annuli** *also* **annuluses** a ring-shaped part, structure, or marking

anodyne *n* **1** a drug that eases pain

anoestrus *adj or n* (of) the period in which there is no sexual activity between 2 periods of sexual activity in cyclically breeding mammals (eg dogs)

anorexia nervosa *n* pathological aversion to food induced by emotional disturbance and typically accompanied by emaciation

anosmia *n* (partial) loss of the sense of smell

anovulant *n or adj* (a drug) that suppresses ovulation

anoxia *n* hypoxia, esp so severe that it causes permanent damage

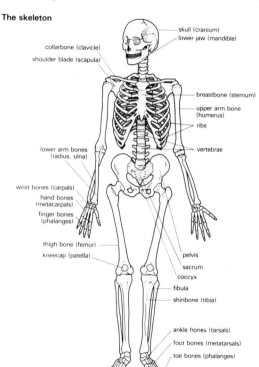

The skeleton

skull (cranium)

lower jaw (mandible)

collarbone (clavicle)

shoulder blade (scapula)

breastbone (sternum)

upper arm bone (humerus)

ribs

vertebrae

lower arm bones (radius, ulna)

wrist bones (carpals)

hand bones (metacarpals)

finger bones (phalanges)

thigh bone (femur)

kneecap (patella)

pelvis

sacrum

coccyx

fibula

shinbone (tibia)

ankle bones (tarsals)

foot bones (metatarsals)

toe bones (phalanges)

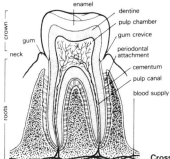

Cross section of a molar tooth

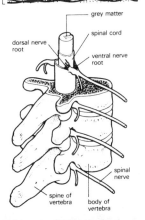

The spine

The spine consists of a flexible column of 7 cervical, 12 thoracic, and 5 lumbar vertebrae separated by cartilaginous discs, together with 5 fused sacral vertebrae and 3 to 5 fused caudal vertebrae which collectively form the sacrum and the coccyx.

ant

antacid *adj* that corrects excessive acidity, esp in the stomach

antagonism *n* opposition in physiological or biochemical action, esp between an agonist and an antagonist

antagonist *n* a drug that opposes the action of another or of a substance (eg a neurotransmitter) that occurs naturally in the body

antagon·ize, -ise *vt* to oppose or counteract

antenna *n, pl* **antennae, antennas** a movable segmented sense organ on the head of insects, myriapods, and crustaceans ☞ INSECT

anterior *adj* **1** situated before or towards the front: eg **a** *of an animal part* near the head; cephalic **b** *of the human body or its parts* ventral **2** *of a plant part* (on the side) facing away from the stem or axis; *also* inferior

anther *n* the part of a stamen that contains and releases pollen ☞ FLOWER

antheridium *n, pl* **antheridia** the male reproductive organ of a fern or related plant

anthocyanin *also* **anthocyan** *n* any of various blue to red plant pigments

anthozoan *n* any of a class of marine coelenterates that includes the corals and sea anemones

anthrax *n* an often fatal infectious disease of warm-blooded animals (eg cattle, sheep, or human beings) caused by a spore-forming bacterium

anthropoid *adj* resembling human beings or the anthropoid apes (eg in form or behaviour); apelike

antibiosis *n* antagonism between organisms, specif microorganisms, or between one organism and a metabolic product of another

antibody *n* a protein (eg an immunoglobulin) that is produced by the body in response to a specific antigen and that counteracts its effects (eg by neutralizing toxins or grouping bacteria into clumps)

anticodon *n* a group of 3 nucleotide bases in a transfer RNA molecule that identifies the amino acid carried and that binds to a complementary codon in messenger RNA during protein synthesis at a ribosome

antigen *n* a protein, carbohydrate, etc that stimulates the production of an antibody when introduced into the body

antihistamine *n* any of various substances that oppose the actions of histamine and are used esp for treating allergies and motion sickness

antipyretic *n or adj* (sthg, esp a drug) that reduces fever

antisepsis *n* the inhibiting of the growth of microorganisms by antiseptic means

antiseptic¹ *adj* **1a** opposing sepsis (in living tissue), specif by arresting the growth of microorganisms, esp bacteria **b** of, acting or protecting like, or using an antiseptic **2** scrupulously clean; aseptic

antiseptic² *n* an antiseptic substance; *also* a germicide

antiserum *n* a serum containing antibodies

antitoxin *n* (a serum containing) an antibody capable of neutralizing the specific toxin that stimulated its production in the body

antrum *n, pl* **antra** the cavity of a hollow organ or sinus

anuran *n or adj* (a) salientian

anus *n* the rear excretory opening of the alimentary canal

anvil *n* the incus

aorta *n, pl* **aortas, aortae** the great artery that carries blood from the left side of the heart to be distributed by branch arteries throughout the body ⟲ RESPIRATION

aortic arch *n* any of a series of paired arterial branches in vertebrate embryos that connect the front and back arterial systems in front of the heart and persist in a complete form only in adult fishes

ape *n* any family of large semierect tailless or short-tailed Old World primates, (eg the gorilla and chimpanzee) ⟲ PRIMATE

apetalous *adj* having no petals

apex *n, pl* **apexes, apices 1** the uppermost peak; the vertex **2** the narrowed or pointed end; the tip **3** the growing tip of a plant

aphasia *n* (partial) loss of the power to use or understand words, usu resulting from brain damage — compare ALEXIA

aphid *n* a greenfly or related small sluggish insect that sucks the juices of plants

apian *adj* of bees

apical dominance *n* inhibition of the growth of lateral buds by the terminal bud or shoot

aplasia *n* incomplete or faulty development of an organ or part

apodal, apodous *adj* having no (appendages analogous to) feet <*eels are* ~>

apoenzyme *n* a protein that forms an active enzyme by combination with a coenzyme

apogamy *n* development of a sporophyte from a gametophyte without fertilization

apomixis *n, pl* **apomixes** reproduction involving the production of seed without fertilization

apophysis *n, pl* **apophyses** an expanded or projecting part, esp of an organism

aposematic *adj, esp of insect coloration* conspicuous and serving to warn

appendage *n* a limb, seta, or other subordinate or derivative body part

appendicitis *n* inflammation of the vermiform appendix

appendix *n, pl* **appendixes, appen-**

13

dices the vermiform appendix or similar bodily outgrowth

apterous *adj* lacking wings <~ *insects*>

aquarium *n, pl* **aquariums, aquaria 1** a glass tank, artificial pond, etc in which living aquatic animals or plants are kept **2** an establishment where collections of living aquatic organisms are exhibited

aquatic *adj* growing living in, or frequenting water

aqueous humour *n* a transparent liquid occupying the space between the lens and the cornea of the eye ⏢ SENSE ORGAN

arable *n or adj* (land) being or fit to be farmed for crops

arachnid *n* any of a class (eg spiders, mites, ticks, and scorpions) of arthropods whose bodies have 2 segments of which the front bears 4 pairs of legs

arachnoid *n* a thin membrane covering the brain and spinal cord and lying between the dura mater and the pia mater

arboreal *adj* of, resembling, inhabiting, or frequenting a tree or trees

arborescent *adj* resembling a tree in properties, growth, structure, or appearance

arbor·ize, -ise *vi* to assume a tree-like appearance <*the nerve fibres* ~d>

archegonium *n, pl* **archegonia** the flask-shaped female sex organ of mosses, ferns, and some conifers

archetype *n* an inherited idea or mode of thought derived from the collective unconscious

arenaceous *adj* growing or living in sandy places

arginine *n* an amino acid that is a chemical base and is found in most proteins

arid *adj* excessively dry; *specif* having insufficient rainfall to support agriculture

aril *n* an exterior covering of some seeds (eg those of yew) that develops from the stalk of the ovule after fertilization

arithmetic mean *n*, a value calculated by dividing the sum of a set of terms by the number of terms

armature *n* an offensive or defensive structure in a plant or animal (eg teeth or thorns)

arrest *n* the condition of being stopped, checked, or made inactive <*cardiac* ~>

arrhythmia *n* an (abnormal) alteration in rhythm of the heartbeat

artefact, artifact *n* sthg (eg a structure seen in the microscope) unnaturally present through extraneous influences (eg from defects in the staining process)

arterial *adj* of or (being the bright red blood) contained in an artery

arteriole *n* a very small artery connecting a larger artery with (small blood vessels like) capillaries

arteriosclerosis *n* abnormal thickening and hardening of the arterial walls

artery *n* any of the branching elastic-walled blood vessels that carry blood away from the heart to the lungs and through the body — compare VEIN

arthritis *n, pl* **arthritides** usu painful inflammation of 1 or more joints

arthropod *n* any of a phylum of invertebrate animals (eg insects, arachnids, and crustaceans) with a jointed body and limbs and usu an outer skin made of chitin that is shed at intervals

articular *adj* of a joint <~ *cartilage*>

articulate[1] *adj* jointed

articulate[2] *vt* to unite by means of a joint ~ *vi* to become united or connected (as if) by a joint

articulation *n* a (movable) joint (between plant or animal parts)

artificial insemination *n* introduction of semen into the uterus or oviduct by other than natural means

artificial respiration *n* the rhythmic forcing of air into and out of the lungs of sby whose breathing has stopped

asbestosis *n, pl* **asbestoses** a disease of the lungs due to the inhalation of asbestos particles

ascarid *n* the common roundworm, parasitic in the human intestine, or a related nematode

ascidian *n* any of an order of tunicates (eg the sea squirt); *broadly* a tunicate

ascomycete *n* any of a class of higher fungi (eg yeast) in which the spores are formed in asci

ascorbic acid *n* VITAMIN C

ascus *n, pl* **asci** the membranous oval or tubular spore sac of an ascomycete

aseptic *adj* 1 preventing infection <~ *techniques*> 2 free or freed from disease-causing microorganisms <*an ~ operating theatre*>

asexual *adj* 1 lacking sex (organs) 2 produced without sexual action or differentiation

asexual reproduction *n* reproduction (eg cell division or spore formation) without union of individuals or germ cells

asparagine *n* an amino acid that is an amide of aspartic acid and that serves as a form of storage for amino groups in plants

aspartic acid *n* an amino acid found in most proteins, esp those of plants

aspergillus *n, pl* **aspergilli** any of a genus of fungi including many common moulds

asphyxia *n* a lack of oxygen in the body, usu caused by interruption of breathing, and resulting in unconsciousness or death

aspirate[1] *vt* to draw or remove (eg blood) by suction

aspirate[2] *n* material removed by aspiration

aspiration *n* a drawing of sthg in, out, up, or through (as if) by suction: eg **a** the act of breathing (sthg in) **b** the withdrawal of fluid from the body

aspirator *n* an apparatus for aspirating (fluid, tissue, etc from the body); *also* a pooter

assimilate *vt* to take in or absorb into the system (as nourishment)

association *n* **1** the formation of mental connections between sensations, ideas, memories, etc **2** an ecological community with usu 2 or more dominant species uniformly distributed

aster *n* a system of cytoplasmic rays typically arranged radially about a centrosome at either end of the mitotic spindle

asthma *n* (an allergic condition marked by attacks of) laboured breathing with wheezing and usu coughing, gasping, and a sense of constriction in the chest

astigmatism *n* a defect of an optical system (eg a lens or the eye) in which rays from a single point fail to meet in a focal point, resulting in a blurred image

asymptomatic *adj* presenting no symptoms of disease

atavism *n* (an individual or character showing) recurrence in (the parts of) an organism of a form typical of ancestors more remote than the parents

ataxia *n* an inability to coordinate voluntary muscular movements that is symptomatic of some nervous disorders

atherosclerosis *n* arteriosclerosis with the deposition of fatty substances in and fibrosis of the inner layer of the arteries

athlete's foot *n* a fungal infection of the feet

Atlantic *adj* of or found near the Atlantic ocean

atlas *n* the first vertebra of the neck

ATP *n* a derivative of adenine that is reversibly converted, esp to ADP, with the release of the cellular energy required for many metabolic reactions [*a*denine *t*ri*p*hosphate]

atrium *n*, *pl* **atria** *also* **atriums** an anatomical cavity or passage; *specif* a chamber of the heart that receives blood from the veins and forces it into a ventricle or ventricles ⤳ RESPIRATION

atrophy[1] *n* (sometimes natural) decrease in size or wasting away of a body part or tissue

atrophy[2] *vb* to (cause to) undergo atrophy

atropine *n* a substance found in deadly nightshade and used in medicine to inhibit the parasympathetic nervous system

attenuate[1] *vt* to reduce the severity, virulence, or vitality of ~ *vi to become thin or fine; diminish*

attenuate[2] *adj* tapering gradually <*an ~ leaf*> ⤳ PLANT

audiology *n* the biology of hearing

audiometer *n* an instrument for measuring the sharpness of hearing

auditory *adj* of or experienced through hearing

aural *adj* of the ear or the sense of hearing

auricle *n* **1a** PINNA **2 b** an atrium of the heart — not now in technical use **2** an ear-shaped lobe

auricular *adj* **1** of or using the ear or the sense of hearing **2** of an auricle

auscultation *n* the act of listening to the heart, lungs, etc as a medical diagnostic aid

australopithecine *adj* of extinct southern African manlike creatures with near-human teeth and a relatively small brain

autism *n* a disorder of childhood development marked esp by inability to form relationships with other people

autoantibody *n* an antibody that combines with a constituent of an individual's own tissues rather than with foreign matter (eg bacteria)

autochthonous *adj* indigenous, native — compare ALLOCHTHONOUS

autogamy *n* self-fertilization

autogenous, autogenic *adj* originating or derived from sources within the same individual <*an ~ graft*>

autograft *n* a transplant from one part to another part of the same body

autoimmune *adj* of or caused by autoantibodies; *specif, of a disease* caused by the production of large numbers of autoantibodies

autointoxication *n* a state of being poisoned by toxic substances produced within the body

autologous *adj* derived from the same individual

autolysis *n* breakdown of all or part of a cell or tissue by self-produced enzymes

autonomic *adj* **1** acting or occurring involuntarily <~ *reflexes*> **2** relating to, affecting, or controlled by the autonomic nervous system

autonomic nervous system *n* a part of the vertebrate nervous system that supplies smooth and cardiac muscle and glandular tissues with nerves and consists of the sympathetic nervous system and the parasympathetic nervous system

autopsy *n* a postmortem examination

autoradiograph, autoradiogram *n* an image produced by radiation from a radioactive substance in an object in close contact with a photographic film or plate

autosome *n* a chromosome other than a sex chromosome

autotrophic *adj* able to live and grow on carbon from carbon dioxide or carbonates and nitrogen from a simple inorganic compound — compare HETEROTROPHIC

auxin *n* (an analogue of) a plant hormone that promotes growth

auxotrophic *adj* requiring a specific growth substance beyond the minimum required for normal metabolism and reproduction <~ *mutants of bacteria*>

aversion therapy *n* therapy intended to change antisocial behaviour or a habit by association with unpleasant sensations

avian *adj* of or derived from birds

avitaminosis *n, pl* **avitaminoses** disease resulting from a deficiency of 1 or more vitamins

awn *n* any of the slender bristles at the end of the flower spikelet in some grasses (eg barley)

axil *n* the angle between a branch or leaf and the axis from which it arises

axillary *adj* 1 of or located near the armpit 2 situated in or growing from an axil

axis *n, pl* **axes** 1 the second vertebra of the neck on which the head and first vertebra pivot 2 any of various parts that are central, fundamental, or that lie on or constitute an axis 3 a plant stem

axon *n* a usu long projecting part of a nerve cell that usu conducts impulses away from the cell body

azoic *adj* having no life; *specif* of the geological time that antedates life

azotobacter *n* any of a genus of large rod-shaped or spherical bacteria that occur in soil and sewage and fix atmospheric nitrogen

azygous, azygos *adj* not being one of a pair <*an ~ vein*>

bacillary *adj* of or caused by bacilli

bacillus *n, pl* **bacilli** a usu rod-shaped bacterium; *esp* one that causes disease

back mutation *n* mutation of a previously mutated gene to its former condition

bacteraemia *n* the usu transient presence of microorganisms, esp bacteria, in the blood

bactericide sthg that kills bacteria

bacteriology *n* 1 a science that deals with bacteria 2 bacterial life and phenomena <*the ~ of a water supply*>

bacteriolysis *n* destruction or dissolution of bacterial cells

bacteriophage *n* any of various specific viruses that attack bacteria

bacteriostasis *n* inhibition of the growth of bacteria without their destruction

bacterium *n, pl* **bacteria** any of a group of microscopic organisms that live in soil, water, organic matter, or the bodies of plants and animals and are important to human beings because of their chemical effects and because many of them cause diseases ⊃ EVOLUTION

bag of waters *n* the double-walled fluid-filled sac that encloses and protects the foetus in the womb and that breaks, releasing its fluid, during the birth process

ball-and-socket joint n a joint (eg in the hip) in which a rounded part moves within a socket so as to allow free movement in many directions

bank n a place where data, human organs, blood, etc are held available for use when needed

barb n 1 any of the side branches of the shaft of a feather ⊃ BIRD 2 a plant hair or bristle ending in a hook

barbel n a slender tactile projecting organ on the lips of certain fishes (eg catfish) used in locating food

barbule n any of the small outgrowths that fringe the barbs of a feather ⊃ BIRD

bark n the tough exterior covering of a woody root or stem

barley n a widely cultivated cereal grass whose seed is used to make malt and in foods (eg breakfast cereals and soups) and stock feeds

barnacle n any of numerous marine crustaceans that are free-swimming as larvae but fixed to rocks or floating objects as adults

barrel n the trunk, esp of a quadruped

barren adj **1a** of a female or mating incapable of producing offspring **b** habitually failing to fruit **2** not productive: eg lacking a normal cover of vegetation or crops; desolate

barrier n a factor that tends to restrict the free movement, mingling, or interbreeding of individuals or populations

basal adj of, situated at, or forming the base

basal metabolic rate n the rate at which heat is given off by an organism at complete rest

basal metabolism n the rate at which energy is used in a fasting and resting organism using energy solely to maintain vital cellular activity, respiration, and circulation

base n **1** that part of an organ by which it is attached to another structure nearer the centre of a living organism **2** a supporting or carrying ingredient **3** a usu water soluble substance that reacts with an acid to form a salt and turns red litmus blue

basidiomycete n any of a large class of higher fungi bearing spores on a basidium and including rusts, mushrooms, and puffballs

basidiospore n a spore produced by a basidium

basidium n, pl **basidia** a specialized cell on a basidiomycete bearing usu 4 basidiospores

basilar adj of or situated at the base

basilar membrane n a membrane in the cochlea of the inner ear that vibrates in response to sound waves

basophil, basophile n a white blood

bas

cell with basophilic granules — compare EOSINOPHIL

basophilic *adj* staining readily with dyes that are chemical bases

bastard *adj* of an inferior or less typical type, stock, or form

bastard wing *n* the projecting part of a bird's wing corresponding to a mammal's thumb and bearing a few short feathers

bat *n* any of an order of nocturnal flying mammals with forelimbs modified to form wings

Batesian mimicry *n* resemblance of a harmless species to another that is protected from predators by repellent qualities (eg unpleasant taste)

batrachian *n* a frog, toad, or other vertebrate amphibian animal

BCG vaccine *n* a vaccine used to protect people against tuberculosis

beak *n* **1** the bill of a bird; *esp* the bill of a bird of prey adapted for striking and tearing **2** any of various rigid projecting mouth structures (eg of a turtle); *also* the long sucking mouth of some insects

bear[1] *n, pl* **bears** *or esp collectively* **bear** any of a family of large heavy mammals that have long shaggy hair and a short tail and feed largely on fruit and insects as well as on flesh

bear[2] *vi* to produce fruit or offspring; yield

bed *n* **1** (plants grown in) a plot of ground, esp in a garden, prepared for plants **2** the bottom of a body of water; *also* an area of sea or lake bottom supporting a heavy growth of a specified organism <*an oyster* ~>

bedsore *n* a sore caused by prolonged pressure on the tissue of a bedridden invalid

bed-wetting *n* involuntary discharge of urine occurring in bed during sleep

beebread *n* bitter yellowish brown pollen (mixed with honey by bees as food)

beehive *n* a hive

beeswax *n* a yellowish plastic substance secreted by bees that is used by them for constructing honeycombs and is used as a wood polish

beetle *n* any of an order of insects that have 4 wings of which the front pair are modified into stiff coverings that protect the back pair at rest

behaviour, *NAm chiefly* **behavior** *n* **1** anything that an organism does involving action and response to stimulation **2** the response of an individual, group, or species to its environment

behaviourism *n* a theory holding that the proper concern of psychology is the objective study of behaviour and that information derived from introspection is not admissible psychological evidence

behaviour therapy *n* therapy intended to change an abnormal behaviour (eg a phobia) by conditioning the patient to respond normally

belemnite *n* a conical pointed fossil shell of any of an order of extinct cephalopod molluscs

bell *n* the corolla of any of many flowers

belladonna *n* atropine

Bence-Jones protein *n* a protein that occurs abnormally in the blood serum and urine in some cancers of the bone marrow, esp multiple myeloma, and occas in other bone diseases

benign *adj, of a tumour* of a mild character; not malignant

benthos *n* organisms that live on or at the bottom of bodies of water

benzodiazepine *n* any of several chemically related synthetic drugs (eg diazepam, chlordiazepoxide, and nitrazepam) widely used as tranquillizers, sedatives, and hypnotics

benzoic acid *n* an organic acid used esp as a food preservative, in medicine, and in organic synthesis

beriberi *n* a deficiency disease marked by degeneration of the nerves and caused by a lack of or inability to assimilate vitamin B₁

berry *n* **1a** a small, pulpy, and usu edible fruit (eg a strawberry or raspberry) **b** a simple fruit (eg a currant, grape, tomato, or banana) with a pulpy or fleshy pericarp —

used technically in botany **2** an egg of a fish or lobster

bestiality *n* bestial behaviour; *specif* sexual relations between a human being and an animal

beta-adrenergic *adj* of or being a beta-receptor <~ *blocking action*>

beta-blocker *n* a drug (eg propranolol) that inhibits the action of adrenalin and similar compounds and is used esp to treat high blood pressure

beta-oxidation *n* metabolic breakdown of fatty acids, esp in mitochondria

beta-receptor *n* a receptor for neurotransmitters (eg adrenalin) in the sympathetic nervous system whose stimulation is associated esp with dilation of small blood vessels and increased heart rate and output — compare ALPHA-RECEPTOR

B-horizon *n* the subsurface layer of soil that is frequently enriched by substances from the surface layer

bias *n* a tendency of an estimate to deviate in one direction from a true value (eg because of non-random sampling)

biceps *n* the large muscle at the front of the upper arm that bends the arm at the elbow when it contracts; *broadly* any muscle attached in 2 places at one end

bicuspid valve *n* the heart valve consisting of 2 flaps that stops blood flowing back from the left ventricle to the left atrium

bie

biennial *adj, of a plant* growing vegetatively during the first year and flowering, fruiting, and dying during the second

bifid *adj* divided into 2 equal lobes or parts by a central cleft <*a ~ petal*>

biflagellate *adj* having two flagella

bifocal *adj* having 1 part that corrects for near vision and 1 for distant vision <*a ~ lens*>

bifurcate *vi or adj* (to fork so as to be) divided into 2 branches or parts

bilateral symmetry *n* a pattern of symmetry in which the organism is divisible into essentially identical halves by 1 plane only

bile *n* a yellow or greenish fluid secreted by the liver into the intestines to aid the digestion of fats

bilharzia *n* 1 a schistosome 2 schistosomiasis

biliary *adj* of or conveying bile or bile-conveying structures <*~ disorders*>

bilious *adj* marked by or suffering from disordered liver function, esp excessive secretion of bile

bilirubin *n* a reddish yellow pigment occurring in bile, blood, urine, and gallstones

bill *n* (a mouthpart resembling) the jaws of a bird together with variously shaped and coloured horny coverings and often specialized for a particular diet

bilobed *adj* divided into 2 lobes

bilocular *adj* divided into 2 cells or compartments

bimolecular *adj* 1 of or formed from 2 molecules 2 being 2 molecules thick

binary fission *n* asexual reproduction of a cell by division into 2 parts

binaural *adj* of or used with both ears

binocular *adj* of, using, or adapted to the use of both eyes <*good ~ vision*>

binominal *adj, of taxonomic nomenclature* consisting of or using 2 Latin names

binucleate *adj* having 2 nuclei

bioassay *n* the determination of the relative strength of a substance (eg a drug) by comparing its effect on a test organism with that of a standard preparation

biochemistry *n* chemistry that deals with the chemical compounds and processes occurring in organisms

biodegradable *adj* capable of being broken down, esp into simpler harmless products, by the action of living beings (eg microorganisms)

bioenergetics *n pl but sing in constr* the biology of energy transformations and exchanges within and between living things and their environments

bioengineering *n* the application to biological or medical science of engineering principles or equipment

22

biofeedback *n* the technique of making unconscious or involuntary bodily processes perceptible to the senses in order to affect them by conscious mental control

biogenesis *n* 1 the development of living things from preexisting living things 2 biosynthesis

biogenic *adj* produced by living organisms

biogeographical, biogeographic *adj* of or being a geographical region viewed in terms of its plants and animals

biological clock *n* the inherent timing mechanism responsible for various cyclic responses (eg changes in hormone levels) of living beings

biological control *n* control of pests by interference with their ecological environment

biological oxygen demand *n* the amount of oxygen required by microorganisms in water, that can be used as an indicator of pollution

biological warfare *n* warfare involving the use of (disease-causing) living organisms against men, animals, or plants, or the use of chemicals harmful to plants

biology *n* 1 a science that deals with the structure, function, development, distribution, and life processes of living organisms 2a the plant and animal life of a region or environment b the biology of an organism or group

bioluminescence *n* (the emission of) light from living organisms

biomass *n* the amount of living matter present in a region (eg in a unit area or volume of habitat)

biome *n* a major type of ecological community *<the grassland ~>*

biometrics *n pl but sing or pl in constr* biometry

biometry *n* the statistical analysis of biological observations and phenomena

bionic *adj* involving bionics; *also* having or being a bionically designed part (eg a limb)

bionics *n pl but sing or pl in constr* 1 a science concerned with the application of biological systems to engineering problems 2 the use of mechanical parts to replace or simulate damaged parts of a living thing

bionomics *n pl but sing in constr* ecology

biophysics *n pl but sing or pl in constr* a field of study concerned with the application of physics to biological problems

biopsy *n* the removal and examination of tissue, cells, or fluids from the living body

biorhythm *n* a supposed periodic fluctuation in the activity of the biological processes of a living thing that is held to affect and determine mood, behaviour, and performance — usu pl

biosphere *n* 1 the part of the world

in which life exists **2** living beings together with their environment

biosynthesis *n, pl* **biosyntheses** the production of a chemical compound by a living organism

biota *n* the flora and fauna of a region

biotic *adj* of life; *esp* caused or produced by living organisms

biotin *n* a growth-controlling vitamin of the vitamin B complex found esp in yeast, liver, and egg yolk

biotype *n* the organisms sharing a specified genotype

biovular *adj* of fraternal twins derived from two ova

bipartite *adj* cleft (almost) into 2 parts <*a ~ leaf*>

biped *n* a 2-footed animal

bipinnate *adj* doubly pinnate

bird *n* any of a class of warm-blooded vertebrates with the body more or less completely covered with feathers and the forelimbs modified as wings ◉

bird of prey *n* a hawk, vulture, or other bird that feeds on carrion or on meat taken by hunting

birth *n* **1** the emergence of a new individual from the body of its parent **2** the act or process of bringing forth young from within the body

birth control *n* control of the number of children born, esp by preventing or lessening the frequency of conception; *broadly* contraception

birthrate *n* the number of (live) births per unit of population (eg 1000 people) in a period of time (eg 1 year)

bisexual *adj* **1a** possessing characteristics of both sexes **b** sexually attracted to both sexes **2** of or involving both sexes

bitter *adj* being or inducing an acrid, astringent, or disagreeable taste similar to that of quinine that is one of the 4 basic taste sensations — compare SALT, SOUR, SWEET ⟿ SENSE ORGAN

bivalent[1] *adj* **1** having a valence of 2 **2** *of chromosomes* that become associated in pairs during meiotic cell division

bivalent[2] *n* a pair of homologous chromosomes in meiosis

bivalve *n or adj* (a mollusc) having a shell composed of 2 valves ⟿ EVOLUTION

blackfly *n, pl* **blackflies**, *esp collectively* **blackfly** (an infestation by) any of several small dark-coloured insects

blackout *n* a temporary loss or dulling of vision, consciousness, or memory

blackwater fever *n* a severe form of malaria in which the urine becomes dark-coloured

bladder *n* **1** a membranous sac in animals that serves as the recepta-

bird ◉

Parts of a bird: a pigeon

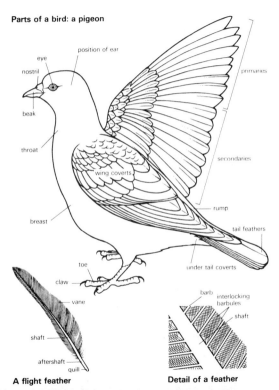

- position of ear
- eye
- nostril
- beak
- throat
- wing coverts
- breast
- primaries
- secondaries
- rump
- tail feathers
- under tail coverts
- toe
- claw
- vane
- shaft
- aftershaft
- quill
- barb
- interlocking barbules
- shaft

A flight feather

Detail of a feather

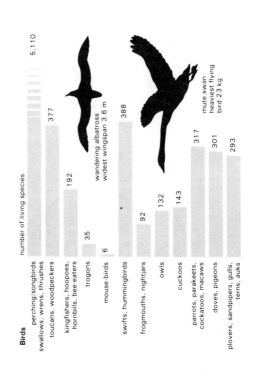

Birds

number of living species

perching/songbirds swallows, wrens, thrushes	5,110
toucans, woodpeckers	377
kingfishers, hoopoes, hornbills, bee-eaters	192
trogons	35
mouse-birds	6
swifts, hummingbirds	388
frogmouths, nightjars	92
owls	132
cuckoos	143
parrots, parakeets, cockatoos, macaws	317
doves, pigeons	301
plovers, sandpipers, gulls, terns, auks	293

wandering albatross
widest wingspan 3.6 m

mute swan
heaviest flying
bird 23 kg

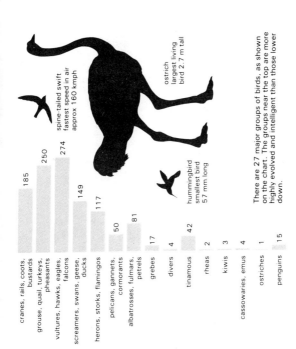

185 cranes, rails, coots, bustards

250 grouse, quail, turkeys, pheasants

274 vultures, hawks, eagles, falcons

spine-tailed swift
fastest speed in air
approx 160 kmph

ostrich
largest living
bird 2.7 m tall

149 screamers, swans, geese, ducks

117 herons, storks, flamingos

50 pelicans, gannets, cormorants

81 albatrosses, fulmars, petrels

hummingbird
smallest bird
57 mm long

17 grebes

4 divers

42 tinamous

2 rheas

3 kiwis

4 cassowaries, emus

1 ostriches

15 penguins

There are 27 major groups of birds, as shown on the chart. The groups near the top are more highly evolved and intelligent than those lower down.

cle of a liquid or contains gas; *esp* the urinary bladder **2** VESICLE 1

blade *n* **1** (the flat expanded part, as distinguished from the stalk, of) a leaf, esp of a grass, cereal, etc **2** a broad flat body part; *specif* the scapula — used chiefly in naming cuts of meat

blast *vi* to shrivel, wither ~ *vt* to injure (as if) by the action of wind; blight

blastema *n* a mass of living substance capable of growth and differentiation

blastocoel *n* the cavity of a blastula

blastocyst *n* the modified blastula of a mammal

blastomere *n* a cell produced during cleavage of an egg

blastula *n, pl* **blastulas, blastulae** the embryo of a metazoan animal at the stage in its development succeeding the morula, typically having the form of a hollow fluid-filled cavity bounded by a single layer of cells — compare GASTRULA, MORULA

bleed¹ *vb* **bled** *vi* **1** to emit or lose blood **2** to lose some constituent (eg sap or dye) by exuding it or by diffusion~ *vt* **1** to remove or draw blood from **2** to draw sap from (a tree)

bleed² *n* an act or instance of bleeding, esp by a haemophiliac

blight¹ *n* (an organism that causes) a disease or injury of plants resulting in withering, cessation of growth, and death of parts without rotting

blight² *vt* to affect (eg a plant) with blight

blind spot *n* the point in the retina where the optic nerve enters that is not sensitive to light ☞ SENSE ORGAN

blindworm *n* a slowworm

blister *n* **1** a raised part of the outer skin containing watery liquid **2** a disease of plants marked by large swollen patches on the leaves

block¹ *n* interruption of the normal physiological function (eg transmission of nerve impulses) of a tissue or organ

block² *vt* to prevent normal functioning of

blood *n* **1** the usu red fluid that circulates in the heart, arteries, capillaries, and veins of a vertebrate animal, carrying nourishment and oxygen to, and bringing away waste products from, all parts of the body **2** a comparable fluid of an invertebrate animal

blood count *n* (the determination of) the number of blood cells in a definite volume of blood

blood group *n* any of the classes into which human beings can be separated on the basis of the presence or absence of specific antigens

blood heat *n* a temperature approximating to that of the human body; about 37°C or 98°F

bloodline *n* a group of related individuals, esp with distinctive characteristics

blood platelet *n* any of the minute cytoplasmic discs in the blood of vertebrates that assist in blood clotting and are non-nucleate in humans

blood poisoning *n* septicaemia

blood pressure *n* pressure that is exerted by the blood on the walls of the blood vessels, esp arteries, and that varies with the age and health of the individual

blood serum *n* the watery portion of the blood excluding the blood cells; *also* blood plasma from which the fibrin has been removed

bloodstream *n* the flowing blood in a circulatory system

blood sugar *n* (the concentration of) the glucose in the blood

blood vessel *n* any of the vessels through which blood circulates in an animal

bloom¹ *n* **1a** a flower **b** the flowering state <*the roses in* ~> **c** an excessive growth of phytoplankton **2** a delicate powdery coating on some fruits and leaves

bloom² *vi* **1** to produce or yield flowers **2** to support abundant plant life <*make the desert* ~>

blowhole *n* **1** a nostril in the top of the head of a whale, porpoise, or dolphin **2** a hole in the ice to which aquatic mammals (eg seals) come to breathe

blubber *n* the fat of large marine mammals, esp whales

blue baby *n* a baby with a bluish tint, usu from a congenital heart defect

blue-green alga *n* any of a class of algae that have no definite nuclei in the cells and a bluish green pigment, phycocyanin

boa *n* any of a number of large snakes (eg the boa constrictor, anaconda, or python) that crush their prey

boar *n* **1a** an uncastrated male pig **1b** the male of several mammals (eg a guinea pig or badger) **2** the Old World wild pig from which most domestic pigs derive

body *n* **1a(1)** the organized physical substance of a living animal or plant **(2)** a corpse **b** a human being; a person **2a** the main part of a plant or animal body, esp as distinguished from limbs and head **b** the main, central, or principal part **3** a group of individuals organized for some purpose **4** compactness or firmness of texture

body cavity *n* a hollow or space within the body, esp the coelom

bog *n* (an area of) wet spongy poorly-drained usu acid ground surrounding a body of open water, with a characteristic flora

boil *n* a localized pus-filled swelling of the skin resulting from infection in a skin gland

bol

bole n the trunk of a tree

boll n the seed pod of cotton or similar plants

boll weevil n a weevil that infests the cotton plant

bolt vi to produce seed prematurely

bolus n **1** a large pill **2** a soft mass of food that has been chewed but not swallowed

bone n (any of the hard body structures composed of) the largely calcium-containing connective tissue of which the adult skeleton of most vertebrate animals is chiefly composed

bony fish n any of a major group of fishes comprising all those with a bony rather than a cartilaginous skeleton and including the salmon, carp, herring, etc; a teleost ⫧ EVOLUTION

book lung n a saclike breathing organ in many arachnids containing numerous thin folds of membrane arranged like the leaves of a book

boreal adj, often cap of or growing in northern and mountainous parts of the northern hemisphere

boscage also **boskage** n a growth of trees or shrubs

botany n **1** a branch of biology dealing with plant life **2a** the plant life (of a region) **b** the properties and life phenomena exhibited by a plant, plant type, or plant group

botfly n any of various heavy-bodied 2-winged flies with larvae parasitic in the alimentary canals of human beings and other large mammals

botulism n acute often fatal food poisoning caused by a toxin in (preserved) food

bovine adj of oxen or cows

bowel n (a specified division of) the intestine or gut — usu pl with sing meaning

Bowman's capsule n the thin membranous capsule surrounding each glomerulus in the kidneys of vertebrates

brachial adj of or located in (a part like) an arm <a ~ artery>

brachiopod n any of a phylum of mostly extinct marine invertebrate animals with shells composed of 2 halves hinged together

brackish adj slightly salty <~ water>

bract n **1** a usu small leaf near a flower or floral axis **2** a leaf borne on a floral axis

bracteole n a small or secondary bract, esp on a floral axis

bradycardia n relatively slow heart action, whether physiological or pathological — compare TACHYCARDIA

brain n **1** the portion of the vertebrate central nervous system that constitutes the organ of thought and neural coordination, is made up of neurons and supporting and nutritive structures, is enclosed within the skull, and is continuous with the spinal cord **2** a nervous centre in invertebrates comparable in position and function to the vertebrate brain

brain death *n* the death of a human being determined by the assessment that his/her brain has irreversibly ceased to function

brain stem *n* the part of the brain connecting the spinal cord with the forebrain and cerebrum

brain wave *n* a rhythmic fluctuation of voltage between parts of the brain

branchia *n, pl* **branchiae** GILL 1

branchiopod *n* any of a group of aquatic crustaceans (eg a brine shrimp) typically having a long body, a carapace, and many pairs of leaflike appendages

brassica *n* any of a large genus of Old World temperate-zone plants of the mustard family that includes many important vegetables and crop plants (eg cabbage, turnip, mustard, and rape)

breakdown *n* **1** a failure to function **2** a physical, mental, or nervous collapse **3** the process of decomposing <~ *of food during digestion*>

breast *n* **1** either of 2 protuberant milk-producing glandular organs situated on the front of the chest in the human female and some other mammals; *broadly* a discrete mammary gland **2** the fore part of the

body between the neck and the abdomen

breastbone *n* the sternum

breastplate *n* the plastron

breathe *vi* **1** to draw air into and expel it from the lungs **2** to live

breech birth *n* a birth in which the rear end of the baby appears first

breed[1] *vb* **bred** *vt* **1** to produce (offspring) by hatching or gestation **2** to propagate (plants or animals) sexually and usu under controlled conditions ~ *vi* **1** to produce offspring by sexual union **2** to propagate animals or plants

breed[2] *n* a group of animals or plants, often specially selected, visibly similar in most characteristics

breeding ground *n* a place or set of circumstances favourable to the propagation or reproduction of an organism

brine *n* water (almost) saturated with common salt

bristle *n* a short stiff coarse hair or filament

bristletail *n* any of 2 orders of wingless insects with 2 or 3 slender bristles at the hind end of the body
☞ INSECT

broad-leaved *adj* having broad leaves; *specif, of a tree* not coniferous

broad-spectrum *adj* effective against a range of insects or microorganisms <*a ~ antibiotic*>

bronchial *adj* of the bronchi or their ramifications in the lungs

31

bro

bronchiectasis *n* abnormal dilation of the bronchial tubes, often as a result of infection

bronchiole *n* a minute thin-walled branch of a bronchus ☞ RESPIRATION

bronchitis *n* (a disease marked by) acute or chronic inflammation of the bronchial tubes accompanied by a cough and catarrh

bronchopneumonia *n* pneumonia involving many widely scattered but small patches of lung tissue

bronchus *n*, *pl* **bronchi** either of the 2 main branches of the windpipe ☞ RESPIRATION

brood[1] *n* young birds, insects, etc hatched or cared for at one time

brood[2] *vi*, *of a bird* to sit on eggs in order to hatch them

broth *n* a liquid medium for culturing esp bacteria

brown alga *n* any of many algae, with a predominantly brown colour, that are mostly seaweeds

brown fat *n* a heat-producing tissue that is present in significant amounts in hibernating mammals, human infants, and adults acclimatized to cold

Brownian movement *n* the random movement of suspended microscopic particles resulting from the impact of molecules of the surrounding fluid or gas

browse *vi*, *of animals* to nibble at leaves, grass, or other vegetation

brucellosis *n* a serious long-lasting disease, esp of human beings or cattle, caused by a bacterium

bruise *n* 1 an injury involving rupture of small blood vessels and discoloration without a break in the skin 2 an injury to plant tissue involving underlying damage and discoloration without a break in the skin

brush *n* (land covered with) scrub vegetation

bryology *n* a branch of botany that deals with mosses and liverworts

bryophyte *n* any of a division of nonflowering plants comprising the mosses and liverworts ☞ PLANT

bryozoan *n* (any of) a phylum or class of aquatic animals that reproduce by budding and usu form colonies

bubo *n*, *pl* **buboes** an inflamed swelling of a lymph gland, esp in the groin or armpit

bubonic plague *n* plague characterized by the formation of buboes

buccal *adj* of or involving the cheeks or the cavity of the mouth

bud[1] *n* 1 a small protuberance on the stem of a plant that may develop into a flower, leaf, or shoot 2a an incompletely opened flower b an outgrowth of an organism that becomes a new individual

bud[2] *vb* **-dd-** *vi* 1 *of a plant* to put forth buds 2 to develop by way of outgrowth 3 to reproduce asexually by forming and developing

32

buds ~ *vt* **1** to produce or develop from buds **2** to graft a bud into (a plant of another kind), usu in order to propagate a desired variety

buffer[1] *n* (a solution containing) a substance capable in solution of neutralizing both acids and bases and thereby maintaining the original acidity or basicity of the solution

buffer[2] *vt* to add a buffer to (eg a solution); *also* to buffer a solution of (a substance)

bug *n* **1** any of an order, of insects that have sucking mouthparts, forewings thickened at the base, and incomplete metamorphosis and are often regarded as pests **2** any of several insects commonly considered obnoxious **3** a disease-producing germ; *also* a disease caused by it — not used technically

bulb *n* **1** a short stem base of a plant (eg the lily, onion, or hyacinth), with 1 or more buds enclosed in overlapping membranous or fleshy leaves, that is formed underground as a resting stage in the plant's development — compare CORM, TUBER **2** a tuber, corm, or other fleshy structure resembling a bulb in appearance **3** a plant having or developing from a bulb **4** a rounded or swollen anatomical structure

bulbil *n* a small or secondary bulb, esp an aerial deciduous bud in a leaf axil

bulbous *adj* **1** growing from or bearing bulbs **2** resembling a bulb, esp in roundness

bulimia *n* an abnormal and constant craving for food

bulk *n* roughage

bundle *n* **1** a small band of mostly parallel nerves or other fibres **2** VASCULAR BUNDLE

bunt *n* a disease of wheat caused by either of 2 parasitic fungi

burr, bur *n* **1** a rough or prickly covering of a fruit or seed **2** sthg that sticks or clings, esp fruit or seed

burrow *n* a hole or excavation in the ground made by a rabbit, fox, etc for shelter and habitation

bursa *n, pl* **bursas, bursae** a small sac or pouch (between a tendon and a bone)

bursitis *n* inflammation of a bursa of the knee, shoulder, elbow, or other joint

bush *n* **1a** a (low densely branched) shrub **b** a close thicket of shrubs **2** a large uncleared or sparsely settled area (eg in Africa or Australia), usu scrub-covered or forested

butterfat *n* the natural fat of milk and chief constituent of butter

butterfly *n* any of numerous slender-bodied day-flying insects with large broad often brightly coloured wings

by-product *n* sthg produced (eg in a chemical reaction) in addition to a principal product

caecum, *NAm chiefly* **cecum** *n* a cavity open at 1 end; *esp* the pouch in which the large intestine begins and into which the ileum opens

caesarean, caesarean section, cae-sarian, *NAm* **cesarean** *n* a surgical incision of the abdominal and ut-erine walls for the delivery of offspring

cainozoic, cenozoic *adj or n* (of or being) an era of geological history that extends from the beginning of the Tertiary period to the present

calcareous *adj* growing on lime-stone or in soil impregnated with lime

calciferol *n* VITAMIN D_2

calcium carbonate *n* a compound found in nature as calcite, lime-stone, etc, and in bones and shells

calcium phosphate *n* any of several phosphates that occur naturally in phosphate rock, teeth, and bones and are used in fertilizers and animal feeds

calculus *n, pl* **calculi** *also* **calculuses** an abnormal hard stony mass (eg of cholesterol) in the kidney, gall bladder, or other hollow organ

calix *n, pl* **calices** CALYX 2

call[1] *vi, of an animal* to utter a characteristic note or cry

call[2] *n* 1 the cry of an animal (eg a bird) 2 (an instrument used to produce) an imitation of an ani-mal's cry made to attract the ani-mal

callose *n* a carbohydrate compo-nent of plant cell walls

callous *adj* hardened and thickened

callus *n* 1 a hard thickened area on skin or bark 2 a mass of connective tissue formed round a break in a bone and changed into bone dur-ing healing 3 soft tissue that forms over a cut plant surface 4 a tumour of plant tissue

calorie *also* **calory** *n* 1 the quantity of heat required to raise the tem-perature of 1g of water by 1°C under standard conditions 2 a kilo-calorie; *also* an equivalent unit expressing the energy-producing value of food when oxidized

calyptra *n* 1 the archegonium of a liverwort or moss; *esp* one forming a hood over the capsule 2 a hood or cap-shaped covering of a flower or fruit 3 ROOT CAP

calyx *n, pl* **calyxes, calyces** 1 the outer usu green or leafy part of a flower or floret, consisting of sep-als 2 **calyx, calix** a cuplike animal structure

cambium *n, pl* **cambiums, cambia** a thin layer of cells between the xylem and phloem of most plants that undergoes cell division to form more xylem and phloem

Cambrian *adj* of or being the earli-est geological period of the Palaeo-zoic era

campanulate *adj* bell-shaped

canal *n* a tubular anatomical chan-nel

cancellous *adj, of bone* porous

cancer *n* (a condition marked by) a malignant tumour of potentially unlimited growth

canine¹ *adj* of or resembling a dog or (members of) the family of flesh-eating mammals that includes the dogs, wolves, jackals, and foxes

canine² *n* any of the 4 conical pointed teeth each of which lies between an incisor and the first premolar on each side of the top and bottom jaws

canker *n* **1** an erosive or spreading sore **2** an area of local tissue death in a plant **3** any of various inflammatory animal diseases

cannon bone *n* the leg bone between the hock joint and the fetlock in hoofed mammals

cannula *n, pl* **cannulas, cannulae** a small tube for insertion into a body cavity or duct

canopy *n* the uppermost spreading branched layer of a forest

cap *n* **1** a usu unyielding overlying rock or soil layer **2** the pileus **3** (a patch of distinctively coloured feathers on) the top of a bird's head **4** ROOT CAP

capillarity *n* the elevation or depression of the surface of a liquid in contact with a solid (eg in a fine-bore tube) that depends on the relative attraction of the molecules in the liquid for each other and for those of the solid

capillary¹ *adj* **1** hair-like; *esp having a very small bore* **2** involving, held by, or resulting from surface tension **3** of capillaries or capillarity

capillary² *n* a tube with a very small bore; *esp* any of the smallest blood vessels connecting arteries with veins and forming networks throughout the body

capitate *adj* **1** forming a head **2** abruptly enlarged and globose

capitulum *n, pl* **capitula 1** a rounded or flattened cluster of stalkless flowers, often simulating 1 larger flower ⤳ FLOWER **2** a rounded protuberance of an anatomical part (eg a bone)

capsulate, capsulated *adj* enclosed in a capsule

capsule *n* **1** a membrane or sac enclosing a body part **2** a thin envelope surrounding a microorganism **3** a closed plant receptacle containing spores or seeds

carapace *n* a hard case (eg of chitin) covering (part of) the back of a turtle, crab, etc

carbamide *n* urea

carbohydrate *n* any of various compounds of carbon, hydrogen, and oxygen (eg sugars, starches, and celluloses) formed by green plants and constituting a major class of energy-providing animal foods

carbon *n* a nonmetallic element occurring as diamond, graphite, charcoal, coke, etc that is important as a constituent of organic compounds

carbon 14 *n* a heavy radioactive carbon isotope (of mass number 14) used in carbon dating

carbon cycle *n* the cycle of carbon in living things in which carbon dioxide from the air is converted by photosynthesis to organic substances and is then ultimately restored to the inorganic state by respiration and rotting ☞ ECOLOGY

carbon dating *n* the dating of ancient material (eg an archaeological specimen) by recording the amount of carbon 14 remaining

carbon dioxide *n* a heavy colourless gas that does not support combustion, is formed esp by the combustion and decomposition of organic substances, and is absorbed from the air by plants in photosynthesis

carboniferous *adj* **1** producing or containing carbon or coal **2** *cap* of or being the period of the Palaeozoic era between the Devonian and the Permian during which coal deposits formed

carboxylase *n* an enzyme that catalyses a chemical reaction in which a carboxyl group is added or removed

carboxyl group *n* an acid radical typical of organic acids

carbuncle *n* a painful local inflammation of the skin and deeper tissues with multiple openings for the discharge of pus

carcass, *Br also* **carcase** *n* a dead body; *esp* the prepared body of a meat animal

carcinogen *n* sthg (eg a chemical compound) that causes cancer

carcinoma *n, pl* **carcinomas, carcinomata** a malignant tumour of epithelial origin

cardiac *adj* **1** of, situated near, or acting on the heart **2** of the oesophageal end of the stomach **3** of heart disease

cardiograph *n* an instrument that registers graphically movements of the heart

cardiovascular *adj* of or involving the heart and blood vessels

caries *n* a progressive destruction of bone or tooth; *esp* tooth decay

carnassial *adj* of or being the large long cutting teeth of a carnivore

carnivore *n* a flesh-eating animal; *esp* any of an order of flesh-eating mammals ☞ MAMMAL

carnivorous *adj* **1** of or being a carnivore; *specif* flesh-eating **2** *of a plant* feeding on nutrients obtained from animal tissue, esp insects

carotene *n* any of several orange or red hydrocarbon plant pigments convertible to vitamin A

carotenoid *also* **carotinoid** *n* a carotene or similar animal or plant pigment

carotid *adj or n* (of or being) the

chief artery or pair of arteries that supply the head with blood

carpel *n* any of the structures of a flowering plant that constitute the female (innermost) part of a flower and usu consist of an ovary, style, and stigma ☞ FLOWER

carpophagous *adj* feeding on fruits

carpus *n, pl* **carpi** (the bones of) the wrist

carrier *n* 1 a bearer and transmitter of a causative agent of disease; *esp* one who is immune to the disease 2 a usu inactive accessory substance; VEHICLE 1 3 a substance (eg a catalyst) by whose agency some element or group is transferred from one compound to another

cartilage *n* (a structure composed of) a translucent elastic tissue that makes up most of the skeleton of very young vertebrates and becomes mostly converted into bone in adult higher vertebrates

cartilaginous fish *n* 1 any of a major group of fishes comprising all those with a cartilaginous rather than a bony skeleton and including the sharks, dogfishes, and rays; an elasmobranch ☞ EVOLUTION 2 a cyclostome

caruncle *n* 1 a naked fleshy outgrowth (eg a domestic fowl's wattle) 2 an outgrowth on a seed adjacent to the micropyle

caryopsis *n, pl* **caryopses, caryopsides** a small 1-seeded dry fruit (eg of grasses) in which the fruit and

seed are fused together in a single grain

casein *n* a protein in milk that is precipitated by (lactic) acid or rennet, is the chief constituent of cheese, and is used in making plastics

cast[1] *vt* **cast** 1 to shed, moult 2 *of an animal* to give birth to (prematurely)

cast[2] *n* 1 a slight squint in the eye 2 PLASTER 2 3 the excrement of an earthworm

caste *n* a specialized form of a social insect (eg a soldier or worker ant) adapted to carry out a particular function in the colony

castrate *vt* 1 to remove the testes of; geld 2 to remove the ovaries of; spay

cat *n* any of a family of carnivores that includes the domestic cat, lion, tiger, leopard, jaguar, cougar, lynx, and cheetah

catabolism *n* destructive metabolism involving the breakdown of complex materials (eg glucose) and resulting in the release of energy and) — compare ANABOLISM

catabolite *n* a (waste) product of catabolism

catadromous *adj* living in fresh water and going to the sea to spawn <~ *eels*>

catalepsy *n* a trancelike state (associated with schizophrenia) in which the body remains rigid and immobile for prolonged periods

cat

catalyse, *NAm* **catalyze** *vt* to bring about the catalysis of (a chemical reaction)

catalysis *n*, *pl* **catalyses** a change, esp an increase, in the rate of a chemical reaction induced by a catalyst

catalyst *n* a substance (eg an enzyme) that changes, esp increases, the rate of a chemical reaction but itself remains chemically unchanged

catalytic *adj* causing or involving catalysis

cataplexy *n* sudden temporary paralysis following a strong emotional stimulus (eg shock)

cataract *n* clouding of (the enclosing membrane of) the lens of the eye; *also* the clouded area

catarrh *n* (the mucus resulting from) inflammation of a mucous membrane, esp in the human nose and air passages

catatonia *n* (a psychological disorder, esp schizophrenia, marked by) catalepsy

caterpillar *n* a wormlike larva, specif of a butterfly or moth ➯ INSECT

catharsis *n*, *pl* **catharses** 1 purgation 2 the process of bringing repressed ideas and feelings to consciousness and expressing them, esp during psychoanalysis

catheter *n* a tubular device for insertion into a hollow body part (eg a blood vessel), usu to inject or draw off fluids or to keep a passage open

cation *n* the ion in an electrolyzed solution that migrates to the cathode; *broadly* a positively charged ion

caudal *adj* 1 of or being a tail 2 situated at or directed towards the hind part of the body

caudate *also* **caudated** *adj* having a tail or tail-like appendage

caulescent *adj*, *of a plant* having a stem that shows above the ground

cauline *adj* of or growing on (the upper part of) a stem — compare RADICAL 1

caustic *adj* capable of destroying or eating away by chemical action

cauter·ize, -ise *vt* to sear or destroy (eg a wound or body tissue) with a cautery, esp in order to rid of infection

cautery *n* 1 cauterization 2 an instrument (eg a hot iron) or caustic chemical used to cauterize tissue

cecum *n*, *pl* **ceca** *NAm* the caecum

celiac *adj*, *NAm* coeliac

cell *n* 1 a small compartment (eg in a honeycomb), receptacle, cavity (eg one containing seeds in a plant ovary), or bounded space 2 the smallest structural unit of living matter consisting of nuclear and cytoplasmic material bounded by a semipermeable membrane and capable of functioning either alone or with others in all fundamental life processes

38

cell body *n* the nucleus-containing central part of a neuron excluding its axons and dendrites

cell division *n* the process by which 2 daughter cells are formed from a parent cell — compare MEIOSIS, MITOSIS

cell membrane *n* PLASMA MEMBRANE

cell sap *n* cytoplasm

cellular *adj* of, relating to, or consisting of cells

cellulase *n* an enzyme that hydrolyses cellulose

cellulitis *n* diffuse, esp subcutaneous, inflammation of body tissue

cellulose *n* a polysaccharide of glucose units that constitutes the chief part of plant cell walls, occurs naturally in cotton, kapok, etc, and is the raw material of many manufactured goods (eg paper, rayon, and cellophane)

cell wall *n* the firm nonliving wall, formed usu from cellulose, that encloses and supports most plant cells

cenozoic *n or adj* Cainozoic

centipede *n* any of a class of many-segmented arthropods with each segment bearing 1 pair of legs ➪ EVOLUTION

central nervous system *n* the part of the nervous system which in vertebrates consists of the brain and spinal cord and which coordinates the activity of the entire nervous system

centrifugal force *n* the force that tends to impel (a part of) sthg outward from the centre of rotation

centriole *n* either of a pair of organelles consisting of 9 microtubules arranged cylindrically, which are found in many animal cells and function in the formation of the mitotic apparatus

centromere *n* the point on a chromosome by which it appears to attach to the spindle in mitosis

centrosome *n* (the region of clear cytoplasm that contains) a centriole

centrum *n, pl* **centrums, centra** the body of a vertebra

cephalic *adj* **1** of or relating to the head **2** directed towards or situated on, in, or near the head

cephalochordate *n* any of a subphylum of marine animals related to vertebrates

cephalopod *n* any of a class of tentacled molluscs that includes the squids, cuttlefishes, and octopuses

cephalothorax *n* the united head and thorax of an arachnid or higher crustacean

cercaria *n, pl* **cercariae** a usu tadpole-shaped larval trematode worm produced in a mollusc host by a redia

cereal[1] *adj* of or relating to (the plants that produce) grain

39

cereal² *n* (a grass or other plant yielding) grain suitable for food

cerebellum *n, pl* **cerebellums, cerebella** a large part of the back of the brain which projects outwards and is concerned esp with coordinating muscles and maintaining equilibrium

cerebral *adj* **1** of the brain or the intellect **2** of or being the cerebrum

cerebral cortex *n* the outer layer of grey matter in the brain whose chief function is the coordination of higher nervous activity

cerebral hemisphere *n* either of the 2 hollow convoluted lateral halves of the cerebrum of the brain

cerebral palsy *n* a disability resulting from damage to the brain before or during birth and characterized by speech disturbance and lack of muscular coordination — compare SPASTIC PARALYSIS

cerebrospinal *adj* of the brain and spinal cord

cerebrospinal fluid *n* a liquid like blood serum that is secreted from the blood into the ventricles of the brain

cerebrovascular *adj* of or involving the brain and the blood vessels supplying it

cerebrum *n, pl* **cerebrums, cerebra** the expanded front portion of the brain that in higher mammals overlies the rest of the brain and consists of the 2 cerebral hemispheres

cervical *adj* of a neck or cervix

cervix *n, pl* **cervices, cervixes 1** (the back part of) the neck **2** a constricted portion of an organ or body part; *esp* the narrow outer end of the uterus ⟹ REPRODUCTION

cesarean *also* **cesarian** *n, NAm* a caesarean

cesspit *n* a pit for the disposal of refuse (eg sewage)

cesspool *n* an underground basin for liquid waste (eg household sewage)

cestode *n* any of a subclass of parasitic flatworms including the tapeworms, usu living in the intestines

cetacean *n* any of an order of aquatic, mostly marine, mammals that includes the whales, dolphins, and porpoises ⟹ MAMMAL

chaeta *n, pl* **chaetae** a bristle, seta

chaetognath *n* any of a phylum of small free-swimming marine worms with movable curved bristles on either side of the mouth

chaff *n* **1** the seed coverings and other debris separated from the seed in threshing grain **2** chopped straw, hay, etc used for animal feed

chalaza *n, pl* **chalazae, chalazas 1** either of a pair of spiral bands in the white of a bird's egg that extend from the yolk and are attached to opposite ends of the lining membrane **2** the point at the base of a plant ovule where the seed stalk is attached

chalcid n any of various related and typically minute insects parasitic in the larval state on the larvae or pupae of other insects

chalk n a soft white, grey, or buff limestone composed chiefly of the shells of small marine organisms

chamaephyte n a perennial plant that bears its over-wintering buds just above soil level

chancre n the initial lesion of some diseases, specif syphilis

chancroid n a bacterial venereal disease

change of life n the menopause

chaparral n a (N American) plant community of shrubs or dwarf trees adapted to dry summers and moist winters

character n 1 an inherited characteristic 2 the sum of the distinctive qualities characteristic of a breed, type, etc; the (distinctive) main or essential nature of sthg

characteristic n a distinguishing trait, quality, or property

cheek pouch n a pouch in the cheek of a monkey, hamster, etc for holding food

chela n, pl **chelae** a pincerlike claw of a crustacean (eg a crab) or arachnid (eg a scorpion)

chelate adj resembling or having chelae

chemiluminescence n light (eg bio-luminescence) produced by chemical reaction

chemoreceptor n a sense organ (eg a taste bud) that responds to chemical stimuli

chemotaxis n orientation or movement of an organism in relation to chemical agents

chemotaxonomy n the classification of plants and animals based on similarities and differences in biochemical composition

chemotherapy n the use of chemical agents in the treatment or control of disease

chemotropism n orientation or growth of cells or sedentary organisms in relation to chemical stimuli

chernozem n a dark-coloured humus-rich soil found in temperate to cool climates

chiasma n, pl **chiasmata** an anatomical cross-shaped configuration; esp that between paired chromatids considered to be the point where genetic material is exchanged

chicken pox n an infectious virus disease, esp of children, that is marked by mild fever and a rash of small blisters

chilblain n an inflammatory sore, esp on the feet or hands, caused by exposure to cold

chimaera n 1 any of a family of marine cartilaginous fishes with a tapering tail 2 a chimera

chimera n an individual, organ, or part consisting of tissues of diverse genetic constitution and occurring esp in plants and most frequently at a graft union

chiropody *n* the care and treatment of the human foot in health and disease

chiropractic *n* a system of healing disease that employs manipulation and adjustment of body structures (eg the spinal column)

chiropter, chiropteran *n* a bat ☞ MAMMAL

chi-square, chi-squared *n* a statistic that indicates the agreement between a set of observed values and a set of values derived from a theoretical model

chitin *n* a horny polysaccharide that forms part of the hard outer covering of esp insects and crustaceans

chiton *n* any of an order of marine molluscs with a shell of many plates

chlorophyll *n* **1** the green photosynthetic colouring matter of plants found in the chloroplasts **2** a waxy green chlorophyll-containing substance extracted from green plants and used as a colouring agent or deodorant

chloroplast *n* a chlorophyll-containing organelle that is the site of photosynthesis and starch formation in plant cells

chlorosis *n* **1** an iron-deficiency anaemia of young girls characterized by a greenish colour of the skin **2** a diseased condition in

green plants marked by yellowing or blanching

cholecystokinin *n* a hormone secreted by the lining of the duodenum that regulates the emptying of the gall bladder and secretion of enzymes by the pancreas

cholera *n* (any of several diseases of human beings and domestic animals similar to) an often fatal infectious epidemic disease caused by a bacterium and marked by severe gastrointestinal disorders

cholesterol *n* a hydroxy steroid that is present in animal and plant cells and is a possible factor in hardening of the arteries

cholic acid *n* a bile acid important in fat digestion

choline *n* a naturally occurring substance that is a vitamin of the vitamin B complex essential to liver function

chordate *n or adj* (any) of a phylum or subkingdom of animals including the vertebrates that have at some stage of development a notochord, a central nervous system along the back, and gill clefts

chorion *n* the outer embryonic membrane of higher vertebrates associated with the allantois in forming the placenta

C-horizon *n* the layer of soil lying beneath the B-horizon and consisting of weathered rock

choroid, choroid coat *n* a membrane containing large pigment cells that lies between the retina and the sclera of the vertebrate eye ➤ SENSE-ORGAN

chromatid *n* either of the paired strands of a chromosome

chromatin *n* a complex of DNA with proteins that forms the chromosomes in the cell nucleus and is readily stained

chromatogram *n* the visual record (eg the pattern remaining in the absorbent medium) of the components separated by chromatography

chromatography *n* the separation of chemicals from a mixture by passing the mixture as a solution or vapour over or through a substance (eg paper) which adsorbs the chemicals to differing extents

chromatophore *n* a pigment-bearing cell or organelle: eg **a** any of the cells found in the surface layer of an animal that are capable of causing skin-colour changes by expanding or contracting **b** the organelle of photosynthesis in blue-green algae and photosynthetic bacteria

chromomere *n* any of the small bead-shaped concentrations of chromatin that are arranged in a line along the chromosome

chromophore *n* a chemical group that gives rise to colour in a compound

chromoplast *n* a coloured body in a plant cell that contains no chlorophyll but usu contains red or yellow pigment (eg carotene)

chromoprotein *n* a compound (eg haemoglobin) of a protein with a metal-containing pigment (eg haem)

chromosome *n* any of the gene-carrying bodies that contain DNA and protein and are found in the cell nucleus

chromosome number *n* the usu constant number of chromosomes characteristic of a particular species of animal or plant

chronic *adj* **1** *esp of an illness* marked by long duration or frequent recurrence — compare ACUTE 1 **2** suffering from a chronic disease

chrysalis *n, pl* **chrysalides, chrysalises** (the case enclosing) a pupa, esp of a butterfly or moth

chyle *n* lymph that is milky from emulsified fats and is produced during intestinal absorption of fats

chyme *n* the semifluid mass of partly digested food expelled by the stomach into the duodenum

chymotrypsin *n* an enzyme that breaks down proteins and is released into the intestines from the pancreas during digestion

cicatrix *n, pl* **cicatrices 1** a scar resulting after a flesh wound has healed **2** a mark resembling a scar: eg **a** a mark left on a stem after the fall of a leaf or bract **b** HILUM 1

ciliary *adj* 1 of cilia 2 of or being the ciliary body

ciliary body *n* the ringlike muscular body supporting the lens of the eye ☞ SENSE ORGAN

cilium *n, pl* **cilia** 1 an eyelash 2 a minute hairlike part; *esp* one capable of a lashing movement that produces locomotion in a single-celled organism

circadian *adj* being, having, characterized by, or occurring in approximately day-long periods or cycles (eg of biological activity or function) <~ *rhythms*><~ *leaf movements*>

circinate *adj* rolled or coiled (with the top as a centre) <~ *fern fronds unfolding*>

circulatory system *n* the system of blood, blood and lymphatic vessels, and heart concerned with the circulation of the blood and lymph

circumcise *vt* to cut off the foreskin of (a male) or the clitoris of (a female)

cirrhosis *n, pl* **cirrhoses** hardening (of the liver) caused by excessive formation of connective tissue

cirrus *n, pl* **cirri** 1 a tendril 2 a slender usu flexible (invertebrate) animal appendage

cistron *n* a gene consisting of a segment of DNA which codes for a particular enzyme, RNA molecule, etc

citric acid *n* an acid occurring in lemons, limes, etc, formed as an intermediate in cell metabolism, and used as a flavouring

citric acid cycle *n* KREBS CYCLE

cladistics *n* a theory that describes the relationship between types of organism on the assumption that their sharing of a unique characteristic (eg the hair of mammals) possessed by no other organism indicates their descent from a single common ancestor

cladode *n* a branch that closely resembles a leaf and often bears leaves or flowers

clasper *n* a male copulatory structure of some insects and fishes

class *n* 1 a group, set, or kind sharing common attributes 2 a category in biological classification ranking above the order and below the phylum or division

classification *n* systematic arrangement in groups according to established criteria; *specif* taxonomy

claustrophobia *n* abnormal dread of being in closed or confined spaces

clavate *adj* club-shaped

clavicle *n* a bone of the vertebrate shoulder typically linking the shoulder blade and breastbone; the collarbone ☞ ANATOMY

claviform *adj* club-shaped

claw *n* 1 (a part resembling or limb having) a sharp usu slender curved nail on an animal's toe 2 any of the pincerlike organs on the end of

coc

some limbs of a lobster, scorpion, or similar arthropod

clay *n* (soil composed chiefly of) an earthy material that is soft when moist but hard when fired, is composed mainly of fine particles of aluminium silicates, and is used for making brick, tile, and pottery

clay loam *n* a loam soil containing from 20 to 30 percent clay

cleavage *n* CELL DIVISION

cleft palate *n* a congenital fissure of the roof of the mouth

climacteric[1] *adj* of or being a critical period (eg of life)

climacteric[2] *n* the menopause; *also* a corresponding period in the male during which sexual activity and competence are reduced

climax *n* **1** an orgasm **2** a relatively stable final stage reached by a (plant) community in its ecological development

cline *n* a graded series of differences in shape or physiology shown by a group of related organisms, usu along a line of environmental or geographical transition; *broadly* a continuum

clitellum *n* a saddle-like region of some annelid worms that secretes a sac in which eggs are deposited

clitoris *n* a small erectile organ at the front or top part of the vulva that is a centre of sexual sensation in females ➣ REPRODUCTION

cloaca *n, pl* **cloacae** the chamber into which the intestinal, urinary,

and generative canals discharge, esp in birds, reptiles, amphibians, and many fishes

clone[1] *n* **1** an individual that is asexually produced and is therefore identical to its parent **2** all such progeny of a single parent — used technically

clone[2] *vt* to cause to grow (as if) as a clone

clostridium *n, pl* **clostridia** any of various spore-forming soil or intestinal bacteria that cause gas gangrene, tetanus, and other diseases

clot[1] *n* a coagulated mass produced by clotting of blood

clot[2] *vb* **-tt-** *vi* **1** to become a clot; form clots **2** *of blood* to undergo a sequence of complex chemical and physical reactions that results in conversion from liquid form into a coagulated mass ~ *vt* to cause to clot

club moss *n* any of an order of primitive vascular plants ➣ EVOLUTION, PLANT

coagulant *n* sthg that produces coagulation

coagulate *vb* to (cause to) become viscous or thickened into a coherent mass; curdle, clot

coagulum *n, pl* **coagula, coagulums** a coagulated mass

coccidiosis *n, pl* **coccidioses** a disease of birds (eg poultry) and mammals (eg sheep) caused by coccidia

coccidium *n, pl* **coccidia** any of an

order of protozoans usu parasitic in the lining of the digestive tract of vertebrates

coccus *n, pl* **cocci** a spherical bacterium

coccyx *n, pl* **coccyges** *also* **coccyxes** the end of the spinal column below the sacrum in human beings and the tailless apes ☞ ANATOMY

cochlea *n, pl* **cochleas, cochleae** a coiled part of the inner ear of higher vertebrates that is filled with liquid through which sound waves are transmitted to the auditory nerve ☞ SENSE ORGAN

cocoon *n* (an animal's protective covering similar to) a (silk) envelope which an insect larva forms about itself and in which it passes the pupa stage

code *vt* to specify (an amino acid, protein, etc) in terms of the genetic code ~ *vi* to be or contain the genetic code *for* an amino acid, protein, etc

codon *n* a group of 3 adjacent nucleotides in RNA or DNA that codes for a particular amino acid or starts or stops protein synthesis

coelacanth *n* any of a family of mostly extinct fishes

coelenterate *n* any of a phylum of radially symmetrical invertebrate animals including the corals, sea anemones, and jellyfishes

coeliac, *NAm chiefly* **celiac** *adj* of the abdominal cavity

coelom *n, pl* **coeloms, coelomata** the usu epithelium-lined space between the body wall and the digestive tract in animals more advanced than the lower worms

coelomate *n or adj* (an animal) having a coelom

coenocyte *n* a syncytium

coenzyme *n* a nonprotein compound that combines with a protein to form an active enzyme and whose activity cannot be destroyed by heat

coenzyme A *n* a coenzyme that occurs in all living cells and is essential to the metabolism of carbohydrates, fats, and some amino acids

cofactor *n* a substance that acts with another substance to bring about certain effects; *esp* a coenzyme

cognition *n* (a product of) the act or process of knowing that involves the processing of sensory information and includes perception, awareness, and judgment

cohort *n* a group of individuals having age, class membership, or other statistical factors in common in a study of the population

coitus *n* the natural conveying of semen to the female reproductive tract; *broadly* SEXUAL INTERCOURSE

coitus interruptus *n* coitus which is purposely interrupted in order to prevent ejaculation of sperm into the vagina

colchicine *n* a substance extracted

from the corms or seeds of the meadow saffron and used esp to inhibit division of the cell nucleus in mitosis and in the treatment of gout

cold-blooded *adj* having a body temperature not internally regulated but approximating to that of the environment — compare WARM-BLOODED

coleoptera *n pl* the insects that are beetles

coleopteran *n* a beetle

coleoptile *n* the first leaf produced by a germinating seed of grasses and some related plants, that forms a protective sheath round the bud that develops into the shoot

coleorhiza *n* the protective sheath surrounding the radicle of a germinating grass seed

colic *n* a paroxysm of abdominal pain localized in the intestines or other hollow organ and caused by spasm, obstruction, or twisting

colitis *n* inflammation of the colon

collagen *n* an insoluble protein that occurs as fibres in connective tissue (eg tendons) and in bones and yields gelatin and glue on prolonged heating with water

collapse¹ *vi* to break down in energy, stamina, or self-control through exhaustion or disease; *esp* to fall helpless or unconscious

collapse² *n* **1** an (extreme) breakdown in energy, strength, or self-control **2** an airless state of (part of) a lung

collarbone *n* the clavicle

collective unconscious *n* that part of a person's unconscious which is inherited and shared with all other people

collembolan *n* any of an order of primitive wingless arthropods (eg springtails)

collenchyma *n* a plant tissue of growing stems, leaf midribs, etc that consists of living (elongated) cells with irregularly thickened walls — compare PARENCHYMA, SCLERENCHYMA

colloid *n* **1** a system consisting of a colloid together with the gaseous, liquid, or solid medium in which it is dispersed **2** a gelatinous substance found in tissues

colon *n, pl* **colons, cola** the part of the large intestine that extends from the caecum to the rectum

colonize *vt* **1** to establish a colony in or on **2** to establish in a colony ~ *vi* to make or establish a colony

colony *n* **1** a distinguishable localized population within a species <*a ~ of termites*> **2** a mass of microorganisms, usu growing in or on a solid medium **3** all the units of a compound animal (eg a coral)

colostomy *n* surgical formation of an artificial anus

colostrum *n* the milk that is secreted for a few days after giving birth and is characterized by high protein and antibody content

colour, *NAm chiefly* **color** *n* **1** the visual sensation (eg red or grey) caused by the wavelength of perceived light that enables one to differentiate otherwise identical objects **2** the aspect of objects and light sources that may be described in terms of hue, lightness, and saturation for objects and hue, brightness, and saturation for light sources **3** a hue, esp as opposed to black, white, or grey

colour-blind *adj* (partially) unable to distinguish 1 or more colours

columnar *adj* of, being, or composed of tall narrow (somewhat) cylindrical epithelial cells

coma *n* a state of deep unconsciousness caused by disease, injury, etc

commensalism *n* the association of 2 species whereby one or both species obtain benefits (eg food or protection) without either species being harmed

commissure *n* **1** the place where 2 parts are joined; a closure **2** a connecting band of nerve tissue in the brain or spinal cord

community *n* all the interacting populations of various living organisms in a particular area

compensation *n* **1** increased functioning or development of one organ to compensate for a defect in another **2** the alleviation of feelings of inferiority, frustration, failure, etc in one field by increased endeavour in another

competition *n* active demand by 2 or more (kinds of) organisms for some environmental resource in short supply

complement *n* the protein in blood serum that in combination with antibodies causes the destruction of antigens (eg bacteria)

complementary *adj* of the precise pairing of bases between 2 strands of DNA or RNA such that the sequence of bases on one strand determines that on the other

complex[1] *adj* of 2 or more parts; difficult to separate, analyse, or solve

complex[2] *n* a group of repressed related desires and memories that usu adversely affects personality and behaviour

complication *n* a secondary disease or condition developing in the course of a primary disease

composite *adj* of or belonging to a very large family of plants, including the dandelion, daisy, and sunflower, typically having florets arranged in dense heads that resemble single flowers

compound *adj* composed of or resulting from union of (many similar) separate elements, ingredients, or parts

compound eye *n* an arthropod eye consisting of a number of separate visual units ☞ INSECT

conceive *vt* to become pregnant with (young) ~ *vi* to become pregnant

conception *n* conceiving or being conceived

conchology *n* the branch of zoology that deals with shells

concrescence *n* a growing together; a coalescence

concretion *n* a hard usu inorganic mass formed (abnormally) in a living body

concussion *n* (a jarring injury to the brain often resulting in unconsciousness caused by) a stunning or shattering effect from a hard blow

condition *vt* to modify so that an act or response previously associated with one stimulus becomes associated with another

conditioned *adj, esp of a reflex* determined or established by conditioning

conduction *n* the transmission of an electrical impulse through (nerve) tissue

conductivity *n* the quality or power of conducting or transmitting

condyle *n* a prominence of a bone forming part of a joint

cone *n* **1** a mass of overlapping woody scales that, esp in trees of the pine family, are arranged on an axis and bear seeds between them; *broadly* any of several similar flower or fruit clusters **2** any of the relatively short light receptors in the retina of vertebrates that are

sensitive to bright light and function in colour vision — compare ROD

congenital *adj* existing at or dating from birth <~ *idiocy*>

conidiophore *n* a structure (on a fungal hypha) that bears conidia

conidium *n, pl* **conidia** an asexual spore (eg of a fungus or bacterium)

conifer *n* any of an order of mostly evergreen trees and shrubs including pines, cypresses, and yews, that bear ovules naked on the surface of scales rather than enclosed in an ovary ➔ EVOLUTION, PLANT

conjugant *n* either of a pair of conjugating gametes or organisms

conjugate *vi* **1** to become joined together **2** to pair and fuse in genetic conjugation

conjugated *adj* formed by the combination of 2 compounds or combined with another compound <*a* ~ *protein*>

conjugation *n* **1** fusion of (similar) gametes with union of their nuclei that in algae, fungi, etc replaces the typical fertilization of higher forms **2** the one-way transfer of DNA between bacteria in cellular contact

conjunctiva *n, pl* **conjunctivas, conjunctivae** the mucous membrane that lines the inner surface of the eyelids and is continued over part of the eyeball

connective tissue *n* any of various tissues (eg bone or cartilage) that pervade, support, and bind together other tissues and organs

conoid, conoidal *adj* shaped (nearly) like a cone

conscious *adj* having mental faculties undulled by sleep, faintness, or stupor; awake

consciousness *n* **1** the totality of conscious states of an individual **2** the upper level of mental life of which sby is aware, as contrasted with unconscious processes

conservation *n* careful preservation and protection, esp of a natural resource, the quality of the environment, or plant or animal species, to prevent exploitation, destruction, etc

conserve *vt* to avoid wasteful or destructive use of <~ *natural resources*>

consociation *n* an ecological community with a single dominant organism

constipation *n* abnormally delayed or infrequent passage of faeces

constrictor *n* **1** a muscle that contracts a cavity or orifice or compresses an organ **2** a snake (eg a boa constrictor) that kills prey by compressing it in its coils

consumer *n* an organism requiring complex organic compounds for food, which it obtains by preying on other organisms or by eating particles of organic matter ⟹ ECOLOGY

contact inhibition *n* the cessation

of movement and growth of one cell when in contact with another, observed esp in tissue cultures

contagious *adj* communicable by contact; catching

contingency table *n* a table that shows the correlation between 2 variables

contraception *n* prevention of conception or impregnation

contraceptive *n* a method or device used in preventing conception;

contractile *adj* having the power or property of contracting <*a ~ protein*>

contractile vacuole *n* a vacuole in a protozoan organism that contracts to discharge fluid from the body

contraction *n* the shortening and thickening of a muscle (fibre)

contracture *n* a permanent shortening of muscle, tendon, scar tissue, etc producing deformity

control *n* (an organism, culture, etc used in) an experiment in which the procedure or agent under test in a parallel experiment is omitted and which is used as a standard of comparison in judging experimental effects

convergence, convergency *n* **1** a converging, esp towards union or uniformity; *esp* coordinated movement of the eyes resulting in reception of an image on corresponding retinal areas **2** independent development in unrelated organisms of similar characters, often associated with similar environments or behaviour

convolution *n* any of the irregular ridges on the surface of the brain, esp of the cerebrum of higher mammals

convulse *vt* to shake or agitate violently, esp (as if) with irregular spasms

convulsion *n* **1** an abnormal violent and involuntary contraction or series of contractions of the muscles **2** an uncontrolled fit; a paroxysm

copepod *n* any of a large subclass of usu minute freshwater and marine crustaceans

coprolite *n* fossil excrement

coprophagous *adj* feeding on dung <*a ~ beetle*>

copulate *vi* to engage in sexual intercourse

copulation *n* sexual intercourse

coral *n* (the hard esp red deposit produced as a skeleton chiefly by) a colony of anthozoan polyps

coralline *n* **1** any of a family of hardened calcium-containing red seaweeds **2** any of various aquatic invertebrate animals, specif a bryozoan or hydroid, that live in colonies and resemble coral

cord *n* an anatomical structure (eg a nerve) resembling a cord

cordate *adj* heart-shaped ☞ PLANT

corepressor *n* a substance that activates a particular genetic repressor (eg by combining with it)

cork *n* **1** the elastic tough outer tissue of the cork oak used esp for stoppers and insulation **2** the phellem of a plant

corm *n* a rounded thick underground plant stem base with buds and scaly leaves — compare BULB, TUBER

cornea *n* the transparent part of the coat of the eyeball that covers the iris and pupil ☞ SENSE ORGAN

corolla *n* the petals of a flower constituting the inner floral envelope

corona *n* **1** the upper portion of a bodily part (eg a tooth or the skull) **2** a circular appendage on the inner side of the corolla in the daffodil, jonquil, etc

coronary *adj* (of or being the arteries or veins) of the heart

coronary artery *n* either of 2 arteries that supply the tissues of the heart with oxygenated blood

coronary thrombosis *n* the blocking of a coronary artery of the heart by a blood clot, usu causing death of heart muscle tissue

coronary vein *n* any of several veins that drain blood from the tissues of the heart

corpus callosum *n, pl* **corpora callosa** a wide band of nerve fibres joining the cerebral hemispheres in the brains of humans and other higher mammals

corpuscle *n* **1** a living (blood) cell **2**

any of various very small multicellular parts of an organism

corpus luteum n, pl **corpora lutea** a reddish-yellow mass of hormone-secreting tissue that forms in the mammalian ovary after ovulation and quickly returns to its original state if the ovum is not fertilized

correlation n 1 reciprocal relation in the occurrence of different structures, characteristics, or processes in organisms 2 an interdependence between mathematical variables, esp in statistics 3 a relation of phenomena as invariable accompaniments of each other

cortex n, pl **cortices, cortexes** 1 the outer part of the kidney, adrenal gland, a hair, etc; esp the outer layer of grey matter of the brain 2 the layer of (parenchymatous) tissue between the inner vascular tissue and the outer epidermal tissue of a green plant

cortical adj 1 (consisting) of a cortex 2 involving or resulting from the action or condition of the cerebral cortex

corticate adj having a cortex

corticosteroid n any of various adrenal-cortex steroids

corticotrophin n ADRENOCORTICO-TROPHIC HORMONE

cortisol n hydrocortisone

corymb n a flat-topped inflorescence; specif one in which the flower stalks are attached at differ-

ent levels on the main axis ⟳ FLOWER

coryza n short-lasting infectious inflammation of the upper respiratory tract; esp a common cold

cosmopolitan adj, of a plant, animal, etc found in most parts of the world and under varied ecological conditions

cotyledon n 1 a lobule of the placenta of a mammal 2 the first leaf or either of the first pair or whorl of leaves developed by the embryo of a seed plant

couple vi **coupling** to copulate

cover glass n a piece of very thin glass used to cover material on a glass microscope slide

covert n a feather covering the bases of the wing or tail feathers of a bird ⟳ BIRD

covey n a mature bird or pair of birds with a brood of young; also a small flock

cowfish n a sirenian

cowpox n a mild disease of the cow that when communicated to humans gives protection against smallpox

coxa n, pl **coxae** the basal segment of a limb of an insect, spider, etc ⟳ INSECT

crab louse n a sucking louse that infests the pubic region of the human body

cramp n 1 a painful involuntary spasmodic contraction of a muscle 2 pl severe abdominal pain

cranial nerve *n* any of the (12 pairs of) nerves that leave the lower surface of the brain to connect with the body, esp the head and face

craniate *n or adj* (one) having a skull

craniology *n* a science dealing with variations in size, shape, and proportions of the skull among the different races of human beings

cranium *n, pl* **craniums, crania** the skull; *specif* the part that encloses the brain ☞ ANATOMY

creationist *n or adj* (an adherent) of a theory that all forms of life were created simultaneously by God, and did not evolve from earlier forms

crenate, crenated *adj* having the margin cut into rounded scallops <*a ~ leaf*> ☞ PLANT

crepitation *n* **1** a crackling sound heard from the lungs that is characteristic of pneumonia **2** a grating sound produced by the fractured ends of a bone moving against each other

crepuscular *adj* active in the twilight <*~ insects*>

cretaceous *adj* **1** of or containing large amounts of chalk **2** *cap* of or being the last period of the Mesozoic era

cretinism *n* (congenital) physical stunting and mental retardation caused by severe deficiency of the thyroid gland in infancy

cribriform *adj* pierced with small holes

cricoid *adj* of or being a ring-shaped cartilage of the larynx

crinoid *n* any of a large class of echinoderms having a cup-shaped body with 5 or more feathery arms

crista *n, pl* **cristae** any of the inwardly projecting folds of the inner membrane of a mitochondrion

Cro-Magnon *n* a tall erect race of human beings known from skeletal remains found chiefly in S France and classified as the same species as recent human beings

crop *n* **1** a pouched enlargement of the gullet of many birds in which food is stored and prepared for digestion **2** (the total production of) a plant or animal product that can be grown and harvested extensively <*a large apple ~*>

cross¹ *n* **1** the crossing of dissimilar individuals; *also* the resulting hybrid **2** sby who or sthg that combines characteristics of 2 different types or individuals

cross² *vt* to cause (an animal or plant) to interbreed with one of a different kind; hybridize ~ *vi* to interbreed, hybridize

cross³ *adj* crossbred, hybrid

crossbred *adj* hybrid; *specif* produced by interbreeding 2 pure but different breeds, strains, or varieties

crossbreed¹ *vb* **crossbred** *vt* to hybridize or cross (esp 2 varieties or breeds of the same species) ~ *vi* to undergo crossbreeding

crossbreed[2] *n* a hybrid

cross-fertil·ization, -isation *n* **1** fertilization by the joining of ova with pollen or sperm from a different individual — compare SELF-FER-TILIZATION **2** cross-pollination

crossing-over *n* the interchange of (segments of) genes between homologous chromosomes during meiotic cell division

cross-pollination *n* the transfer of pollen from one flower to the stigma of another — compare SELF-POLLINATION

cross-resistance *n* tolerance (eg of bacteria) to a normally poisonous substance (eg an antibiotic) acquired by exposure to a chemically related substance

crown *n* **1** the topmost part of the skull or head **2** the upper part of the foliage of a tree or shrub **3** (an artificial substitute for) the part of a tooth visible outside the gum ⟳ ANATOMY **4** the part of a flowering plant at which stem and root merge

crucifer *n* any plant of the mustard family, including the cabbage, stock, cress, etc

crumb *n* **1** (loose crumbly soil or other material resembling) the soft part of bread inside the crust **2** a small lump consisting of soil particles

crustacean *n, pl* **crustaceans, crus-**

tacea any of a large class of mostly aquatic arthropods with a carapace, a pair of appendages on each segment, and 2 pairs of antennae, including the lobsters, crabs, woodlice, etc ⟳ EVOLUTION

crustaceous *adj* of, having, or forming a crust or shell; of or being a crustacean

cryptic *adj* serving to conceal <~ *coloration in animals*>

cryptogam *n* a plant (eg a fern, moss, or fungus) reproducing by means of spores and not producing flowers or seed

cryptogenic *adj* of obscure or unknown origin <*a* ~ *disease*>

crystalline lens *n* the lens of the eye in vertebrates

ctenoid *adj* (having or consisting of scales) with a toothed margin <*a* ~ *fish*>

ctenophore *n* any of a phylum of sea animals superficially resembling jellyfishes but swimming by means of 8 bands of thin flat cilia-bearing plates

cuckoo spit *n* (a frothy secretion exuded on plants by the larva of) a froghopper

cucurbit *n* a plant of the cucumber family

cud *n* food brought up into the mouth by a ruminating animal from its first stomach to be chewed again

culm *n* the stem of a grass or other monocotyledonous plant

cultivar *n* an organism of a kind originating and kept under cultivation

cultivate *vt* **1** to prepare or use (land, soil, etc) for the growing of crops; *also* to break up the soil about (growing plants) **2** to foster the growth of (a plant or crop) **3** CULTURE 2

culture[1] *n* **1** cultivation, tillage **2** (a product of) the cultivation of living cells, tissue, viruses, etc in prepared nutrient media

culture[2] *vt* **1** to cultivate **2** to grow (bacteria, viruses, etc) in a culture **3** to start a culture from <~ *a specimen of urine*>

cuneate *adj* having a narrow triangular shape with the smallest angle towards the base <*a ~ leaf*> ☞ PLANT

cuneiform *adj* wedge-shaped

cupule *n* a cup-shaped anatomical structure

curare *n* an extract from a South American vine used as an arrow poison and in medicine as a muscle relaxant

cure[1] *n* (a drug, treatment, etc that gives) relief or esp recovery from a disease

cure[2] *vt* **1** to restore to health, soundness, or normality **2** to bring about recovery from **2** ~ *vt* to effect a cure

curl *n* a (plant disease marked by the) rolling or curling of leaves

cursorial *adj*, *of (a part of) an animal* adapted to running

curvature *n* **1** an abnormal curving (eg of the spine) **2** a curved surface of an organ (eg the stomach)

curve *n* a graphical representation of a variable affected by conditions

Cushing's syndrome *n* obesity, esp of the face, and muscular weakness caused by an excess of glucocorticoid hormones (eg cortisone) often resulting from prolonged therapeutic administration

cusp *n* **1** a point on the grinding surface of a tooth **2** a fold or flap of a heart valve

cutaneous *adj* of or affecting the skin

cuticle *n* a skin or outer covering: eg **a** the (dead or horny) epidermis of an animal **b** a thin fatty film on the external surface of many higher plants

cutis *n*, *pl* **cutes, cutises** the dermis

cutting *n* **1** a part of a plant stem, leaf, root, etc capable of developing into a new plant **2** a harvest

cyanocobalamin *also* **cyanocobalamine** *n* VITAMIN B$_{12}$

cyanophyte *n* BLUE-GREEN ALGA

cycad *n* any of an order of tropical gymnospermous trees resembling palms ☞ PLANT

cyclic AMP *n* a nucleotide in each molecule of which a phosphate group is joined at 2 places to an adenosine group, and which functions as a regulator of processes

cyclosis *n* the slow, usu circular, movement of cytoplasm within a living cell

cyclostome *n or adj* (any) of a class of primitive fishlike vertebrates comprising the hagfishes and lampreys

cyclothymia *n* a condition marked by abnormal swings between elated and depressed moods

cyme *n* an inflorescence in which all floral axes end in a single flower (and the main axis bears the central and first-opening flower with subsequent flowers developing from side shoots) ☞ FLOWER

cypsela *n* an achene with 2 carpels and adherent calyx tube

cyst *n* **1** a closed sac (eg of watery liquid or gas) with a distinct membrane, developing (abnormally) in a plant or animal **2a** (a capsule formed about) a microorganism in a resting or spore stage **b** a resistant cover about a parasite when inside the host

cysteine *n* a sulphur-containing amino acid found in many proteins and readily convertible to cystine

cystic *adj* **1** (composed) of or containing a cyst or cysts **2** of the urinary or gall bladder

cystic fibrosis *n* a common often fatal hereditary disease appearing in early childhood and marked esp by faulty digestion and difficulty in breathing

cystine *n* a sulphur-containing amino acid found in many proteins

cystitis *n* inflammation of the urinary bladder

cytidine *n* a nucleoside containing cytosine

cytochrome *n* any of several enzymes that function in intracellular energy generation as transporters of electrons, esp to oxygen, by undergoing successive oxidation and reduction

cytokinesis *n* the cleavage of the cytoplasm of a cell into daughter cells following division of the nucleus

cytology *n* the biology of (the structure, function, multiplication, pathology, etc of) cells

cytoplasm *n* the substance of a plant or animal cell excluding the organelles (eg the nucleus and mitochondria)

cytosine *n* a pyrimidine base that is one of the 4 bases whose order in a DNA or RNA chain codes genetic information — compare ADENINE, GUANINE, THYMINE, URACIL

cytosol *n* the cytoplasm

cytotoxin *n* a substance with a toxic effect on cells

Darwinian *adj* of (the theories or followers of) Charles Darwin, or Darwinism

Darwinism *n* a theory of evolution asserting that all the groups of plants and animals have arisen by natural selection

data *n pl but sing or pl in constr* factual information (eg measurements or statistics) used as a basis for reasoning, discussion, or calculation

daughter *adj* of the first generation of offspring, molecules, etc produced by reproduction, division, or replication <the ~ cells>

DDT *n* a synthetic chlorinated water-insoluble insecticide that tends to accumulate in food chains and is poisonous to many vertebrates

deaf-mute *n or adj* (one who is) deaf and dumb

deaminate *vt* to remove the amino group from (a compound)

death *n* a permanent cessation of all vital functions; the end of life

decapod *n* **1** any of an order of crustaceans including the shrimps, lobsters, and crabs that have stalked eyes, 5 pairs of appendages, and the head and thorax fused and covered by a carapace **2** any of an order of cephalopod molluscs including the cuttlefishes, squids, etc that have 10 arms

decarboxylase *n* any of a group of enzymes that accelerate decarboxylation, esp of amino acids

decarboxylation *n* the removal of a carboxyl group from a compound

decay[1] *vi* **1** to decline in health, strength, or vigour **2** to undergo

decomposition ~ *vt* to destroy by decomposition

decay[2] *n* **1** (a product of) rot; *specif* decomposition of organic matter (eg proteins), chiefly by bacteria in the presence of oxygen **2** a decline in health or vigour

decidua *n, pl* **deciduae** a part of the lining of the womb that in women and other higher mammals undergoes special changes in preparation for pregnancy and is cast off during menstruation or while giving birth

deciduous *adj* (having parts) that fall off or are shed seasonally or at a particular stage in development <~ teeth><a ~ tree>

decompose *vt* to rot

decomposer *n* any of various organisms (eg fungi or bacteria) that feed on and break down dead plant and animal material ➔ ECOLOGY

decondition *vt* to cause extinction of (a conditioned response)

decumbent *adj, of a plant* lying on the ground except for a raised apex or extremity

decussate *adj, of leaves* arranged in pairs each at right angles to the next pair above or below ➔ PLANT

decussation *n* a crossed tract of nerve fibres passing between parts of the body on opposite sides of the brain or spinal cord; a commissure

defaecate, *also NAm* **defecate** *vb* to

def

discharge (esp faeces) from the bowels

defective *n* one who is subnormal physically or mentally

defence mechanism *n* 1 a reaction by an organism serving to prevent or protect it from attack 2 an (unconscious) mental process (eg projection or repression) that prevents the entry of unacceptable or painful thoughts into consciousness

deficiency *n* a shortage of substances necessary to health

deficiency disease *n* a disease (eg scurvy) caused by a lack of essential vitamins, minerals, etc in the diet

definitive host *n* the host in which the sexual reproduction of a parasite takes place

deflation *n* the erosion of soil by the wind

defoliant *n* a chemical applied to plants to cause the leaves to drop off prematurely

deformity *n* a physical blemish or distortion; a disfigurement

degenerate¹ *adj* having declined in nature, character, structure, function, etc from an ancestral or former state

degenerate² *vi* to evolve or develop into a less autonomous or complex form <~d *into parasites*>

degrade *vt* 1 to erode 2 to decompose

dehisce *vi* to split (open); *also* to

discharge contents by so splitting <*anthers* dehiscing *at maturity*>

dehydrate *vt* to remove (bound) water from (a chemical compound, foods, etc) ~ *vi* to lose water or body fluids (abnormally)

dehydrogenase *n* an enzyme that accelerates the oxidation of or removal of hydrogen from a compound

delirium *n* 1 confusion, frenzy, disordered speech, hallucinations, etc occurring as a (temporary) mental disturbance

delirium tremens *n* a violent delirium with tremors induced by chronic alcoholism

delusion *n* 1 deluding or being deluded 2a sthg delusively believed b (a mental state characterized by) a false belief (about the self or others) that persists despite the facts and occurs esp in psychotic states

dementia *n* 1 deteriorated mentality due to damage to or (natural) deterioration of the brain <*senile* ~> 2 madness, insanity

dementia praecox *n* schizophrenia

demersal *adj* of or living near the bottom of the sea — compare PELAGIC

demography *n* the statistical study of human populations, esp with reference to size and density, distribution, and vital statistics

demyelinate *vt* to remove or destroy the myelin of (a nerve fibre)

58

denature vt to modify (eg a protein) by heat, acid, etc so that some of the original structure of the molecule is lost and its properties are changed

dendrite n any of the (branching) extensions from a nerve cell that conduct impulses towards the body of the cell

dendrochronology n the dating of events and variations in climate by comparative study of the annual growth rings in wood

dendroid adj resembling a tree in form

dendrology n the study of trees

denitrification n the process of denitrifying

denitrify vt to remove (a compound of) nitrogen from; to convert the nitrogen in (a nitrate or nitrite) to gaseous nitrogen released into the atmosphere

denizen n a naturalized plant or animal

density n 1 the quantity per unit volume, unit area, or unit length 2 the average number of individuals or units per unit of space <a population ~>

dentate, dentated adj having teeth or pointed conical projections <a ~ leaf> ⇨ PLANT

dentine n a calcium-containing material, similar to but harder and denser than bone, of which the principle mass of tooth is composed ⇨ ANATOMY

dentition n the number, kind, and arrangement of teeth

deoxygenate vt to remove oxygen from

deoxyribonucleic acid n DNA

deoxyribonucleotide n any of several nucleotides that contain deoxyribose and some of which are constituents of DNA

deoxyribose n a pentose sugar occurring esp in deoxyribonucleotides

dependence also **dependance** n compulsive physiological or psychological need for a habit-forming drug

dependent adj determined or conditioned by another

dependent variable n a variable whose value is determined by the function of another factor —compare DETERMINANT

depopulate vt to reduce greatly the population of ~ vi to decrease in population

depressant n sthg (eg a drug) that depresses function or activity <alcohol acts as a ~ of the brain>

depressed adj lowered or sunken, esp in the centre

depression n 1 (a mental disorder marked by inactivity, difficulty in thinking and concentration, and esp by) sadness or dejection 2 a lowering of activity, vitality, amount, force, etc 3 a depressed place or part

depressor n 1 a muscle that draws

der

down a part — compare LEVATOR **2** a nerve that decreases the activity of an organ or part

derepress *vt* to activate (a gene) by releasing from a blocked state

derm *n* **1** the dermis **2** SKIN 2

dermatitis *n* a disease or inflammation of the skin

dermatology *n* a branch of medicine dealing with (diseases of) the skin

dermis *n* (the sensitive vascular inner layer of) the skin

desalinate *vt* to remove salt from (esp sea water)

desensit·ize, -ise *vt* to make (sby previously sensitive) insensitive or nonreactive to a sensitizing agent

desert *n* (a desolate region like) a dry barren region incapable of supporting much life

desmid *n* any of numerous single-celled or colonial green algae

determinant *n* **1** a gene **2** a factor or element that fixes or conditions an outcome —compare DEPENDENT VARIABLE

detoxify *vt* to remove a poison or toxin from

detritus *n, pl* **detritus** loose material (eg rock fragments or organic particles) produced by disintegration

detumescence *n* subsidence or diminution of swelling

develop *vi* to go through a process of natural growth, differentiation, or evolution by successive changes

development *n* the act, process, or result of developing

deviant[1] *adj* characterized by deviation

deviant[2] *n* a person whose behaviour differs markedly from the norm

deviation *n* **1** the difference between a value in a frequency distribution and a fixed number **2** departure from accepted norms of behaviour <*sexual* ~>

dextrorotatory *adj* turning clockwise or towards the right; *esp* rotating the plane of polarization of light towards the right <~ *crystals*> — compare LAEVOROTATORY

dextrose *n* dextrorotatory glucose

diabetes *n* any of various abnormal conditions characterized by the secretion and excretion of excessive amounts of urine; *specif* DIABETES MELLITUS

diabetes insipidus *n* a disorder of the pituitary gland characterized by intense thirst and by the excretion of large amounts of urine

diabetes mellitus *n* a disorder of the process by which the body uses sugars and other carbohydrates in which not enough insulin is produced or the cells become resistant to its action and which is characterized typically by abnormally great amounts of sugar in the blood and urine

diabetic *n* a person affected with diabetes

diadelphous *adj* united so as to form 2 sets <*the stamens of leguminous plants are* ~>

diakinesis *n* the final stage of the meiotic prophase marked by contraction of the bivalents

dialyse, NAm dialyze *vi* to undergo dialysis

dialysis *n, pl* **dialyses** the separation of substances in solution by means of their unequal diffusion through semipermeable membranes; *esp* the purification of blood by such means

diandrous *adj* having 2 stamens

diapause *n* a period (eg in an insect) of arrested development between periods of activity

diaphragm *n* **1** the partition separating the chest and abdominal cavities in mammals ☞ RESPIRATION **2** a partition in a plant or the body or shell of an invertebrate animal

diarrhoea, NAm chiefly diarrhea *n* abnormally frequent intestinal evacuations with more or less fluid faeces

diastasis *n, pl* **diastases** the rest phase of cardiac diastole occurring between the filling of the ventricle and the emptying of the atrium

diastole *n* a rhythmically recurrent expansion; *esp* the dilation of the cavities of the heart during which they fill with blood — compare SYSTOLE

diatom *n* any of a class of minute single-celled algae with hard shell-like skeletons that are composed of silica

dichotomous *adj* **1** dividing into 2 parts **2** of, involving, or arising from dichotomy

dichotomy *n* a (repeated) branching (into 2 branches)

dichromatic *adj* having 2 colour varieties or colour phases independent of age or sex <*a* ~ *bird*>

dicot *n* a dicotyledon

dicotyledon *n* a plant with 2 seed leaves

dicrotic *adj, of the pulse* having a double beat (eg in certain feverish states)

dictyosome *n* GOLGI BODY

diecious *adj, NAm* dioecious

diet *n* the kind and amount of food habitually or generally consumed by an animal

dietary[1] *n* the kinds and amounts of food available to or eaten by an individual, group, or population

dietary[2] *adj* of (the rules of) a diet

dietetics *n pl but sing or pl in constr* the application of the principles of nutrition to feeding

differentiate *vt* to cause differentiation of in the course of development ~ *vi* to undergo differentiation

differentiation *n* **1** modification of body parts for performance of particular functions **2** the sum of the processes whereby general cells and tissues become specialized and attain their adult form and function

diffuse *vi* to undergo diffusion

diffusion *n* the process whereby particles of liquids, gases, or solids intermingle as the result of their spontaneous movement and in dissolved substances move from a region of high concentration to one of low concentration

digest *vt* to convert (food) into a form the body can use ~ *vi* to become digested

digestion *n* the process or power of digesting sthg, esp food

digestive *adj* of, causing, or promoting digestion

digit *n* a finger or toe

digital *adj* of or with the fingers or toes

digitate *adj* 1 having fingers or toes 2 having divisions arranged like the fingers of a hand <*a ~ leaf*> ☞ PLANT

dihybrid *n or adj* (an organism, cell, etc) being heterozygous for 2 given genes

dilate *vt* to distend ~ *vi* to become wide

dilation *n* the stretching or enlarging of an organ or other part of the body

diluent *n* a diluting agent

dilute *vt* 1 to make thinner or more liquid by adding liquid 2 to diminish the strength, flavour, or brilliance of by adding liquid

dimerous *adj, of an insect or plant part* consisting of 2 parts

dimorphism *n* the occurrence, combination, or existence of 2 distinct forms: eg **a** the existence of 2 different forms of a species, distinguished by size, colour, etc **b** the existence of an organ (eg the leaves of a plant) in 2 different forms **c** crystallization of a chemical compound in 2 different forms

dinoflagellate *n* any of an order of usu marine plankton

dinosaur *n* any of a group of extinct, typically very large flesh- or plant-eating reptiles, most of which lived on the land; *broadly* any large extinct reptile

dinucleotide *n* a nucleotide consisting of 2 units each composed of ribose or deoxyribose combined with a phosphate group and a nitrogen-containing base

dioecious, *NAm also* **diecious** *adj* having male and female reproductive organs in different individuals or plants — compare MONOECIOUS

dipeptide *n* a peptide having 2 molecules of amino acid in its molecular structure

diphtheria *n* an acute infectious disease caused by a bacterium and marked by fever and the formation of a false membrane, esp in the throat, causing difficulty in breathing

diploblastic *adj* having two germ layers — used of an embryo or lower invertebrate that lacks a true mesoderm

diplococcus *n, pl* **diplococci** any of a genus of bacteria that occur usu in pairs and includes some serious pathogens

diploid *n or adj* (a cell or organism) having double the basic number of chromosomes arranged in homologous pairs — compare HAPLOID, POLYPLOID

diplopod *n* a millipede

diplotene *n* a stage of meiotic prophase following pachytene during which the paired chromosomes begin to separate and chiasmata become visible

dipteran *n* TWO-WINGED FLY

dipterous *adj* **1** having 2 wings or winglike appendages **2** of the two-winged flies

disaccharide *n* any of a class of sugars (eg sucrose) that, on hydrolysis, yield 2 monosaccharide molecules

disc, *NAm chiefly* **disk** *n* any of various round flat anatomical structures; *esp* any of the cartilaginous discs between the spinal vertebrae <suffering from a slipped ~ >

disease *n* a condition of (a part of) a living animal or plant body that impairs the performance of a vital function; (a) sickness, malady

disfunction *n* dysfunction

disinfect *vt* to cleanse of infection, esp by destroying harmful microorganisms

disinfectant *n* a chemical that destroys harmful microorganisms

disinfest *vt* to rid of insects, rodents, or other pests

dislocate *vt* to put out of place; *esp* to displace (eg a bone or joint) from normal connection

dislocation *n* displacement of 1 or more bones at a joint

disperse *vt* to cause to become spread widely ~ *vi* to become dispersed or scattered

dispersion *n* **1** the act or process of dispersing **2** the extent to which the values of a frequency distribution are scattered around an average

dissect *vt* to cut (eg an animal or plant) into pieces, esp for scientific examination

dissociation *n* the separation of a more or less autonomous group of ideas or activities from the mainstream of consciousness, esp in cases of mental disorder

distal *adj, esp of an anatomical part* far from the centre or point of attachment or origin; terminal — compare PROXIMAL

distichous *adj* **1** arranged in 2 vertical rows <~ *leaves*> ☞ PLANT **2** divided into 2 segments <~ *antennae*>

distribution *n* **1** the position, arrangement, or frequency of occurrence (eg of the members of a group) over a usu specified area or length of time **2** the natural geographical range of an organism **3**

an arrangement of statistical data that shows the frequency of occurrence of the values of a variable

diuresis *n* an increased excretion of urine

diuretic *n or adj* (a drug) acting to increase the flow of urine

diurnal *adj* **1** having a daily cycle **2** opening during the day and closing at night <~ *flowers*>

divergence *also* **divergency** *n* the acquisition of dissimilar characteristics by related organisms living in different environments

division *n* a group of organisms forming part of a larger group; *specif* a primary category of the plant kingdom equivalent to a phylum of the animal kingdom

dizygotic *also* **dizygous** *adj, of twins* formed from 2 separate zygotes; fraternal

DNA *n* any of various nucleic acids that are found esp in cell nuclei, are constructed as a double helix held together by hydrogen bonds between purine and pyrimidine bases which project inwards from 2 chains containing alternate links of deoxyribose and phosphate, and are responsible for transmitting genetic information

dominant[1] *adj* **1** being the one of a pair of bodily structures that is the more effective or predominant in action <*the* ~ *eye*> **2** being the one of a pair of (genes determining) contrasting inherited character-

istics that predominates — compare RECESSIVE

dominant[2] *n* a socially dominant individual

donor *n* sby used as a source of biological material <*a blood* ~>

dormant *adj* marked by a suspension of activity: eg (appearing to be) asleep or inactive, esp throughout winter

dorsal *adj* relating to or situated near or on the back or top surface esp of an animal or of any of its parts — compare VENTRAL 2

dorsal fin *n* a medium longitudinal vertical fin on the back of a fish or other aquatic vertebrate

double *adj, of a plant or flower* having more than the normal number of petals or ray flowers — compare SINGLE

double-blind *adj* of or being an experimental procedure which is designed to eliminate false results, in which neither the subjects nor the experimenters know the make-up of the test groups and control groups during the actual course of the experiments — compare SINGLE-BLIND

double helix *n* two parallel helices arranged round the same axis; *specif* this arrangement of 2 complementary DNA strands with the bases of each strand pointing inwards and hydrogen-bonding with those of the other, that is regarded as the basic structure of the DNA of most living things

Down's syndrome *n* a form of congenital mental deficiency in which a child is born with slanting eyes, a broad short skull, and broad hands with short fingers; mongolism

drip *n* **1** a device for the administration of a liquid at a slow rate, esp into a vein **2** a substance administered by means of a drip *<a saline ~>*

drive *n* a motivating instinctual need or acquired desire *<a sexual ~>*

drone *n* the male of a bee (eg the honeybee) that has no sting and gathers no honey

dropsy *n* oedema

drosophila *n* any of a genus of small 2-winged fruit flies extensively used in genetic research

drug[1] *n* **1** a substance used as (or in the preparation of) a medication **2** a substance that causes addiction or habituation

drug[2] *vt* **-gg- 1** to affect or adulterate with a drug **2** to administer a drug to **3** to lull or stupefy (as if) with a drug

drum *n* the tympanic membrane of the ear

drupe *n* a fruit (eg a cherry or almond) that has a stone enclosed by a fleshy layer and is covered by a flexible or stiff outermost layer

drupelet *n* a small drupe; *specif* any

of the individual parts of an aggregate fruit (eg the raspberry)

dry rot *n* (a fungus causing) a decay of seasoned timber in which the cellulose of wood is consumed leaving a soft skeleton which is readily reduced to powder

duct *n* **1** a bodily tube or vessel, esp when carrying the secretion of a gland **2** a continuous tube in plant tissue

ductless gland *n* ENDOCRINE GLAND

dung *n* the excrement of an animal

duodenum *n, pl* **duodena, duodenums** the first part of the small intestine extending from the stomach to the jejunum

duplicate *vt* to make double or twofold ~ *vi* to replicate *<DNA in chromosomes ~s>*

dura mater *n* the tough fibrous membrane that envelops the brain and spinal cord

Dutch elm disease *n* a fatal disease of elms caused by a fungus, spread from tree to tree by a beetle, and characterized by yellowing of the foliage and defoliation

dwarf *adj* of a (variety of) plant or animal that is smaller than normal size

dynamics *n pl but sing or pl in constr* a pattern of change or growth *<population ~>*

dysentery *n* any of several infectious diseases characterized by se-

dys

vere diarrhoea, usu with passing of mucus and blood

dysfunction, disfunction *n* impaired or abnormal functioning

dysgenics *n pl but sing in constr* the study of racial degeneration

dyslexia *n* a maldevelopment of reading ability in otherwise normal children due to a neurological disorder

dysmenorrhoea *n* painful menstruation

dyspepsia *n* indigestion

dysphasia *n* loss of or deficiency in the power to use or understand language as a result of injury to or disease of the brain

dysphoria *n* a state of feeling unwell or unhappy — compare EUPHORIA

dysplasia *n* abnormal growth or development of organs, cells, etc

dysuria *n* difficult or painful urination

ear¹ *n* 1 (the external part of) the characteristic vertebrate organ of hearing and equilibrium ⟹ SENSE ORGAN 2 any of various organs capable of detecting vibratory motion 3 the sense or act of hearing 4 sensitivity to musical tone and pitch

ear² *n* the fruiting spike of a cereal, including both the seeds and protective structures

eardrum *n* TYMPANIC MEMBRANE

ecdysis *n, pl* **ecdyses** the moulting or shedding of an outer layer (eg in insects and crustaceans)

echinoderm *n* any of a phylum of radially symmetrical marine animals consisting of the starfishes, sea urchins, and related forms

echolocation *n* the location of distant or invisible objects by means of sound waves reflected back to the sender (eg a bat or submarine) by the objects

echovirus *n* any of a group of viruses found in the gastrointestinal tract and sometimes associated with respiratory ailments and meningitis

eclipse plumage *n* comparatively dull plumage that occurs seasonally in ducks or other birds which adopt a distinct nuptial plumage — compare NUPTIAL PLUMAGE

ecology *n* (a science concerned with) the interrelationship of living organisms and their environments 🌐

ecospecies *n, pl* **ecospecies** a taxonomic species regarded as an ecological unit

ecosphere *n* the parts of the universe habitable by living organisms; *esp* BIOSPHERE

ecosystem *n* a complex consisting of a community and its environment functioning as a reasonably self-sustaining ecological unit in nature

ecotype *n* a group equivalent to a taxonomic subspecies and maintained as a distinct population by ecological and geographical factors

ectoderm n 1 the outer cellular layer of an animal having only 2 germ layers (eg a jellyfish) 2 (a tissue derived from) the outermost of the 3 primary germ layers of an embryo

ectogenous, ectogenic adj, esp of pathogenic bacteria capable of development apart from the host

ectomorph n an individual with a body type characterized by slenderness, angularity and fragility

ectoparasite n a parasite that lives on the exterior of its host — compare ENDOPARASITE

ectopic adj occurring in an abnormal position or in an unusual manner or form <~ heartbeat><~ pregnancy>

ectoplasm n the outer relatively rigid granule-free layer of the cytoplasm of a cell — compare ENDOPLASM

edaphic adj of or influenced by the soil

edaphic climax n an ecological climax resulting from soil conditions

edema n, NAm oedema

edentate n or adj (a sloth, armadillo, or other mammal in the same order) having few or no teeth ☞ MAMMAL

effector n a gland, muscle, or other bodily organ that becomes active in response to stimulation

efferent adj conducting outwards from a part or organ; specif con-

veying nervous impulses to an effector — compare AFFERENT

effloresce vi to burst into flower

efflorescence n 1 the period or state of flowering 2 a redness of the skin; an eruption

effluent n sthg that flows out: eg a an outflowing branch of a main stream or lake b smoke, liquid industrial refuse, sewage, etc discharged into the environment, esp when causing pollution

effuse adj spread out flat without definite form <~ lichens>

effusion n 1 an act of effusing 2 the escape of a fluid from a containing vessel; also the fluid that escapes

egest vt to rid the body of (waste material)

egg n 1 the hard-shelled reproductive body produced by a bird; esp that produced by domestic poultry and used as a food 2 an animal reproductive body consisting of an ovum together with its nutritive and protective envelopes that after fertilization is capable of developing into a new individual 3 an ovum

egg tooth n a prominence on the beak or nose of an unhatched bird or reptile that is used to break through the eggshell

ego n, pl **egos** the one of the 3 divisions of the mind in psychoanalytic theory that serves as the organized conscious mediator between the person and reality, esp in

The carbon cycle

carbon dioxide in the air

animal and plant respiration

photosynthesis

combustion

food for animals

fossil fuels, peat, coal, oil, and natural gas

The nitrogen cycle

nitrogen in the air

lightning

nitrogen-fixing bacteria in root nodules

taken up by roots

plant and animal proteins

dead organisms and faeces

nitrite bacteria

nitrate bacteria

denitrifying bacteria

soil nitrate

ecology

A food web

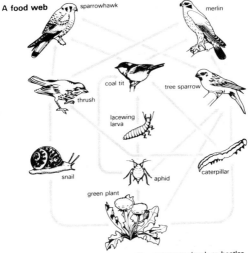

sparrowhawk

merlin

coal tit

tree sparrow

thrush

lacewing larva

snail

aphid

caterpillar

green plant

Dead organisms are consumed by *scavengers* (such as beetles and worms) and *decomposers* (such as bacteria and fungi).

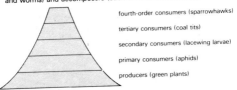

fourth-order consumers (sparrowhawks)

tertiary consumers (coal tits)

secondary consumers (lacewing larvae)

primary consumers (aphids)

producers (green plants)

In the sea, plankton are the producers in the food web.

the perception of and adaptation to reality — compare ID, SUPEREGO

eidetic *adj* marked by or involving extraordinarily accurate and vivid recall of visual images <*an ~ memory*>

ejaculate[1] *vt* to eject from a living body; *specif* to eject (semen) in orgasm

ejaculate[2] *n* the semen released by a single ejaculation

elaborate *vt* to build up (complex organic compounds) from simple ingredients

elasmobranch *n* CARTILAGINOUS FISH 1

elastin *n* a protein similar to collagen that is the chief component of elastic fibres of connective tissue

elbow *n* **1** the joint between the human forearm and upper arm **2** a corresponding joint in the forelimb of a vertebrate animal

Electra complex *n* the Oedipus complex when it occurs in a female

electrocardiogram *n* the tracing made by an electrocardiograph

electrocardiograph *n* an instrument for recording the changes of electrical potential difference occurring during the heartbeat

electroconvulsive therapy *n* a treatment for serious mental disorder, esp severe depression, in which a fit is induced by passing an electric current through the brain

electroencephalogram *n* the tracing made by an electroencephalograph

electroencephalograph *n* an instrument for detecting and recording brain waves

electrolysis *n* the production of chemical changes by the passage of an electric current through an electrolyte

electrolyte *n* a nonmetallic electric conductor often dissolved in a suitable solvent in which current is carried by the movement of ions

electron microscope *n* an electron-optical instrument in which a beam of electrons is used to produce an enlarged image of a minute object

electrophoresis *n* the movement of particles through a gel or other medium in which particles are suspended under the action of an applied electric field

elephantiasis *n, pl* **elephantiases** enormous enlargement of a limb or the scrotum caused by lymphatic obstruction, esp by filarial worms

eliminate *vt* to expel (eg waste) from the living body

elongate[1] *vi* to grow in length

elongate[2], **elongated** *adj* long in proportion to width

elytron *n, pl* **elytra** either of the modified front pair of wings in beetles, cockroaches, and some other insects that serve to protect the hind pair of functional wings

emasculate *vt* to castrate

embolism *n* (the sudden obstruction of a blood vessel by) an embolus

embolus *n, pl* **emboli** a clot, air bubble, or other particle likely to cause an embolism — compare THROMBUS

embryo *n, pl* **embryos** 1 an animal in the early stages of growth before birth or hatching 2 a rudimentary plant within a seed consisting of radicle, plumule and cotyledon(s)

embryology *n* the biology of (the development of) embryos

embryonic *also* **embryonal** *adj* 1 of an embryo 2 in an early stage of development

embryophyte *n* a plant (eg a fern or a flowering plant) that produces an embryo and develops vascular tissues

embryo sac *n* the female part of a female plant consisting of a thin-walled sac containing the egg nucleus, and other nuclei which give rise to endosperm

emulsify *vt* to convert (eg an oil) into an emulsion

emulsion *n* (the state of) a substance (eg fat in milk) consisting of one liquid dispersed in droplets throughout another liquid

enamel *n* a substance composed of calcium phosphate that forms a thin hard layer capping the teeth ⇨ ANATOMY

encapsulate *vt* to enclose (as if) in a capsule ~ *vi* to become encapsulated

encephalic *adj* of the brain

encephalitis *n, pl* **encephalitides** inflammation of the brain, usu caused by infection

encephalogram *n* an X-ray picture of the brain made by encephalography

encephalograph *n* 1 an encephalogram 2 an electroencephalograph

encephalography *n* X-ray photography of the brain after the cerebrospinal fluid has been replaced by a gas (eg air)

encephalomyelitis *n* inflammation of both the brain and spinal cord

encephalon *n, pl* **encephala** the vertebrate brain

encyst *vb* to enclose or become enclosed (as if) in a cyst

endbrain *n* the front subdivision of the forebrain

endemic *adj* belonging or native to a particular area, country, or people <~ disease> <~ species>

endergonic *adj* requiring expenditure of energy <~ *biochemical reactions*>

endocardium *n, pl* **endocardia** a thin membrane lining the cavities of the heart

endocarp *n* the inner layer of the pericarp of a fruit

endocrine *adj* 1 producing secretions that are discharged directly into the bloodstream <~ *system*> — compare EXOCRINE 1 2 of or being an endocrine gland or its secretions <~ *hormone*>

end

endocrine gland *n* the thyroid, pituitary, or other gland that produces an endocrine secretion

endocrinology *n* physiology and medicine dealing with (diseases of) the endocrine glands

endoderm *n* the innermost of the germ layers of an embryo that is the source of the epithelium of the digestive tract and its derivatives

endodermis *n* the innermost layer of cells of the cortex in many roots and stems

endogenous *also* **endogenic** *adj* 1 growing from or on the inside 2 originating within the body

endolymph *n* the watery fluid in the membranous labyrinth of the ear

endometrium *n, pl* **endometria** the mucous membrane lining the uterus

endomorph *n* an individual with a body type characterized by a heavy rounded body build and a tendency to become fat

endoparasite *n* a parasite that lives in the internal organs or tissues of its host — compare ECTOPARASITE

endophyte *n* a plant that lives within another plant

endoplasm *n* the inner relatively fluid part of the cytoplasm of a cell — compare ECTOPLASM

endoskeleton *n* an internal skeleton or supporting framework in an animal

endosperm *n* a nourishing tissue in seed plants that is formed within the embryo sac

endospore *n* an asexual spore developed within a single cell, esp in bacteria

endothelium *n, pl* **endothelia** an inner layer (eg of a seed coat)

endothermic, endothermal *adj* characterized by or formed with absorption of heat

enema *n, pl* **enemas** *also* **enemata 1** injection of liquid into the intestine by way of the anus (eg to ease constipation) **2** material for injection as an enema

energy level *n* any of the divisions of a food chain defined by the method of obtaining food – compare TROPHIC

engorge *vt* to fill (with blood) to the point of congestion ~ *vi, esp of an insect* to suck blood to the limit of body capacity

engulf *vt, of an amoeba, phagocytic cell, etc* to take in (food) by flowing over and enclosing

enteric *adj* of the intestines

enteritis *n* inflammation of the intestines, esp the human ileum, usu marked by diarrhoea

enterokinase *n* an enzyme that converts an inactive substance secreted into the intestines by the pancreas into trypsin

enteron *n* the alimentary canal or system, esp of an embryo

entire *adj* **1** not castrated **2** having

72

the margin continuous or free from indentations <an ~ leaf>

entomology n zoology that deals with insects

entomophagous adj feeding on insects

entomophilous adj being normally pollinated by insects — compare ZOOPHILOUS a

enuresis n an involuntary discharge of urine

environment n the complex of climatic, soil, and biological factors that acts upon an organism or an ecological community

environmentalism n a theory that views environment rather than heredity as the important factor in human development

environmentalist n 1 an advocate of environmentalism 2 sby concerned about the quality of the human environment

enzyme n any of numerous complex proteins that are produced by living cells and catalyse specific biochemical reactions at body temperatures

enzymology n science that deals with enzymes, their nature, activity, and significance

Eocene adj relating to the epoch of the Tertiary after the Palaeocene and before the Oligocene

eosin n a reddish-brown dye used as a biological stain for cytoplasmic structures

eosinophil, eosinophile n a white

blood cell with cytoplasmic granules readily stained by eosin — compare BASOPHIL

ephemeral adj lasting 1 day only <an ~ fever>

epicalyx n a whorl of bracts resembling but exterior to the true calyx

epiblast n the outer layer of an embryo at a very early stage in its development

epicardium n, pl **epicardia** the visceral part of the pericardium that closely covers the heart

epicotyl n the portion of the axis of a plant embryo or seedling above the cotyledon

epidemic n or adj (an outbreak of a disease) affecting many individuals within a population, community, or region at the same time <typhoid was ~>

epidemiology n 1 medicine that deals with the incidence, distribution, and control of disease in a population 2 the factors controlling the presence or absence of (a cause of) disease

epidermis n 1 the thin outer epithelial layer of the animal body that is derived from ectoderm and forms in vertebrates an insensitive layer over the dermis 2 any of various covering layers resembling the epidermis 3 a thin surface layer of tissue in higher plants

epididymis n, pl **epididymides** a mass of convoluted tubes at the back of the testis in which sperm is stored ➔ REPRODUCTION

epidural *adj* situated on or administered outside the dura mater < ~ *anaesthesia* > < ~ *structures* >

epigeal, epigeous *adj* growing, remaining, or occurring above the surface of the ground < ~ *germination of plants* > — compare HYPOGEAL

epiglottis *n* a thin plate of flexible cartilage in front of the glottis that folds back over and protects the glottis during swallowing ➔ SENSE ORGAN

epigynous *adj* (having floral organs) attached to the surface of the ovary and appearing to grow from the top of it — compare HYPOGYNOUS, PERIGYNOUS

epilepsy *n* any of various disorders marked by disturbed electrical rhythms of the brain and spinal chord and typically manifested by convulsive attacks often with clouding of consciousness

epilimnion *n* the water above the thermocline of a lake — compare HYPOLIMNION

epiphysis *n, pl* **epiphyses** 1 an end of a long bone 2 PINEAL GLAND

epiphyte *n* a plant that derives its moisture and nutrients from the air and rain and grows on another plant

epistasis *n, pl* **epistases** suppression of the effect of a gene by another

gene that is not an allele of the first gene

epistaxis *n, pl* **epistaxes** a nosebleed

epithelium *n, pl* **epithelia** 1 a membranous cellular tissue that covers a free surface or lines a tube or cavity of an animal body and serves esp to enclose and protect the other parts of the body, to produce secretions and excretions, and to function in assimilation 2 a usu thin layer of cells that lines a cavity or tube of a plant

erectile *adj* 1 capable of being raised to an erect position; *esp, of animal tissue* capable of becoming swollen with blood to bring about the erection of a body part 2 of or involving the erection of the penis

erection *n* (an occurrence in the penis or clitoris of) the dilation with blood and resulting firmness of a previously flaccid body part

ergometer *n* an apparatus for measuring the work performed by a group of muscles

ergonomics *n pl but sing or pl in constr* a science concerned with the relationship between human beings, the machines they use, and the working environment

ergosterol *n* a steroid found esp in yeast, moulds, and ergot that is converted into vitamin D_2 by ultraviolet light

ergot *n* 1 (a fungus bearing) a black or dark purple club-shaped sclerotium that develops in place of the seed of a grass (eg rye) **2** a disease of rye and other cereals caused by an ergot fungus **3** the dried sclerotia of an ergot fungus containing substances used medicinally (eg to treat migraine)

erogenous *also* **erogenic** *adj* of or producing sexual excitement (when stimulated) <~ *zones*>

error *n* the difference between an observed or calculated value and a true value

erysipelas *n* a feverish disease with intense deep red local inflammation of the skin, caused by infection by a streptococcal bacterium

erythroblast *n* a nucleated bone-marrow cell that gives rise to red blood cells

erythrocyte *n* RED BLOOD CELL

essential amino acid *n* an amino acid (eg lysine) that is required in the diet for normal health and growth

establish *vt* to cause (a plant) to grow and multiply in a place where previously absent

estimate *vt* to judge approximately the value, worth, or significance of

estuarine, estuarial *adj* of, living in, or formed in a river estuary <~ *currents*> <~ *animals*>

ethology *n* the scientific study of animal behaviour

etiolate *vt* to bleach and alter the natural development of (a green plant) by excluding sunlight

etiology *n* aetiology

eucaryote *n* a eukaryote

euchromatin *n* the genetically active part of chromatin that is largely composed of genes

eugenics *n* a science dealing with the improvement (eg by control of human mating) of the hereditary qualities of a race or breed

eukaryote, eucaryote *n* an organism composed of 1 or more cells typically with visibly evident nuclei — compare PROKARYOTE

euphoria *n* an (inappropriate) feeling of well-being or elation — compare DYSPHORIA

eustachian tube *n* a tube connecting the middle ear with the pharynx that equalizes air pressure on both sides of the eardrum ☞ SENSE ORGAN

eutherian *adj or n* (of or being) a mammal of a major division comprising those mammals that have placentas

eutrophic *adj, of a body of water* rich in dissolved nutrients (eg phosphates) but often shallow and seasonally deficient in oxygen

evacuate *vt* to discharge from the body as waste ~ *vi* to pass urine or faeces from the body

evergreen[1] *adj* having leaves that remain green and functional through more than 1 growing season — compare DECIDUOUS

evergreen[2] *n* an evergreen plant; *also* a conifer

evert *vt* to turn outwards or inside out

eviscerate *vt* 1 to disembowel 2 to remove an organ from (a patient); *also* to remove the contents of (an organ)

evolution *n* 1 the historical development of a biological group (eg a race or species) 2 a theory that the various types of animals and plants derived from preexisting types and that the distinguishable differences are due to natural selection ⊕

evolve *vt* to produce by natural evolutionary processes ~ *vi* to undergo evolutionary change

excrescence, excrescency *n* an excessive or abnormal outgrowth or enlargement

excreta *n pl* faeces or other waste matter discharged from the body

excrete *vt* to separate and eliminate or discharge (waste) from blood or living tissue

exfoliate *vt* to cast (eg skin or bark) off in scales, layers, etc ~ *vi* 1 to split into or shed scales, layers, surface body cells, etc 2 to come off in a thin piece 3 to grow (as if) by producing or unfolding leaves

exhale *vt* to breathe out

exocrine *adj* 1 producing secretions that are discharged through a duct — compare ENDOCRINE 1 2 of or being an exocrine gland or its secretions

exocrine gland *n* a gland (eg a sweat gland or a kidney) that releases a secretion external to an organ by means of a duct

exodermis *n* a layer of the outer living cortical cells that functions as the epidermis in roots lacking secondary thickening

exogenous *adj* originating from the outside; due to external causes

exoskeleton *n* an external supportive (hard or bony) covering of an animal

expectorant *n or adj* (sthg) that promotes the discharge of mucus from the respiratory tract

expiration *n* the release of air from the lungs through the nose or mouth

expire *vi* 1 to emit the breath 2 to die — *fml* ~ *vt* to breathe out (as if) from the lungs

exsanguinate *vt* to drain of blood

exserted *adj* projecting beyond an enclosing organ or part <~ *anthers*>

extension *n* a straightening of (a joint between the bones of) a limb

extensor *n* a muscle that produces extension

exteroceptive *adj* activated by, relating to, or being stimuli received by an organism from outside

extinct *adj* no longer existing <*an ~ animal*>

extinction *n* 1 making or being extinct or (causing to be) extinguished 2 elimination or reduction of a conditioned response by not reinforcing it

extinguish *vt* to cause extinction of (a conditioned response)

extracellular *adj* situated or occurring outside a cell or the cells of the body <~ *digestion*> <~ *enzymes*>

extrapolate *vt* **1** to use or extend (known data or experience) in order to surmise or work out sthg unknown **2** to predict by extrapolating known data or experience

extrasensory *adj* residing beyond or outside the ordinary physical senses <*instances of* ~ *perception*>

extrinsic *adj* originating from or on the outside

extrinsic factor *n* VITAMIN B₁₂ — compare INTRINSIC FACTOR

extrovert *also* **extravert** *n* one whose attention and interests are directed wholly or predominantly towards what is outside the self — compare ²INTROVERT 2

exudate *n* matter which has oozed out or spread

eye *n* **1a** any of various usu paired organs of sight; *esp* a nearly spherical liquid-filled organ that is lined with a light-sensitive retina and housed in a bony socket in the skull ☞ SENSE ORGAN **b** the visible parts of the eye with its surrounding structures (eg eyelashes and eyebrows) **2a** a (nearly) circular mark (eg on a peacock's tail) **b** an undeveloped bud (eg on a potato) **c** the (differently coloured or marked) centre of a flower

eyeball *n* the capsule of the eye of a vertebrate formed by the sclera and cornea that cover it, together with the structures they contain

eyespot *n* **1** a simple visual organ of pigment or pigmented cells **2** a spot of colour

eyestalk *n* either of the movable stalks bearing an eye at the tip in a crab or related crustacean

facet *n* the external surface of any of the usu many optical elements of the compound eye of an insect or other arthropod

facial *adj* of the face

facilitation *n* the increase in the ease with which an impulse is conducted along a particular nerve, esp resulting from repetition of the impulse

factor *n* a gene

facultative *adj* having a particular type of life or taking place under some environmental conditions but not under others <*a* ~ *parasite*> — compare OBLIGATE

faeces, *NAm chiefly* **feces** *n pl* bodily waste discharged through the anus

falcate *also* **falcated** *adj* hooked or curved like a sickle ☞ PLANT

falciform *adj* falcate

fallopian tube *n, often cap F* either of the pair of tubes conducting the egg from the ovary to the uterus in mammals ☞ REPRODUCTION

fallow *adj, of land* left unsown after ploughing

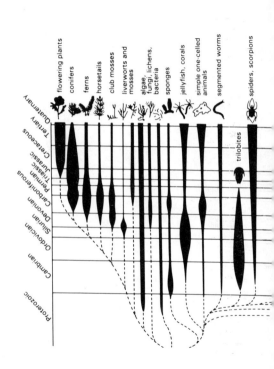

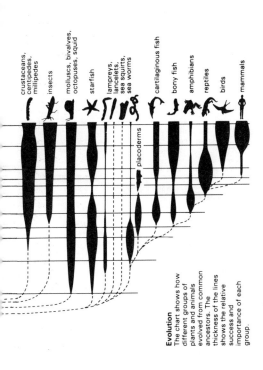

crustaceans, centipedes, millipedes

insects

molluscs, bivalves, octopuses, squid

starfish

lampreys, lancelets, sea squirts, sea worms

placoderms

cartilaginous fish

bony fish

amphibians

reptiles

birds

mammals

Evolution
The chart shows how different groups of plants and animals evolved from common ancestors. The thickness of the lines shows the relative success and importance of each group.

false pregnancy *n* a psychosomatic state in which some of the signs of pregnancy occur without conception

false rib *n* a rib whose cartilages unite indirectly or not at all with the breastbone — compare FLOAT-ING RIB

familial *adj* tending to occur in more members of a family than expected by chance alone <*a ~ disorder*>

family *n* a category in the biological classification of living things ranking above a genus and below an order

family planning *n* a system of achieving planned parenthood by contraception; BIRTH CONTROL

farsighted *adj* of hypermetropica

fascia *n, pl* **fasciae, fascias** 1 a broad well-defined band of colour 2 (a sheet of) connective tissue covering or binding together body structures

fasciation *n* a malformation of plant stems commonly manifested as enlargement and flattening as if several were fused

fascicle *n* a fasciculus

fasciculation *n* muscular twitching in which groups of muscle fibres contract simultaneously

fasciculus *n, pl* **fasciculi** a slender bundle of (anatomical) fibres

fat *n* 1 (animal tissue consisting chiefly of cells distended with greasy or oily matter **2a** any of numerous compounds of carbon, hydrogen, and oxygen that are a major class of energy-rich food and are soluble in organic solvents (eg ether) but not in water **b** a solid or semisolid fat as distinguished from an oil

fathom *n* a unit of length equal to 6ft (about 1.83m) used esp for measuring the depth of water

fatigue *n* 1 physical or nervous exhaustion 2 the temporary loss of power to respond induced in a sensory receptor or motor end organ by continued stimulation

fatty *adj* derived from or chemically related to fat

fatty acid *n* any of numerous organic acids with 1 carboxyl group (eg acetic acid) including many that occur naturally in fats, waxes, and essential oils

fauces *n, pl* **fauces** the narrow passage from the mouth to the pharynx situated between the soft palate and the base of the tongue — often *pl* with sing. meaning

fauna *n, pl* **faunas** *also* **faunae** the animals or animal life of a region, period, or special environment — compare FLORA

feather *n* 1 any of the light horny outgrowths that form the external covering of a bird's body and consist of a shaft that bears 2 sets of barbs that interlock to form a continuous vane ⬅ BIRD

febrifuge *n* an antipyretic

febrile *adj* of fever; feverish

fecund *adj* fruitful in offspring or vegetation; prolific

feed *vb* **fed** *vt* **1** to give food to **2** to give as food **3** to provide sthg essential to the growth, sustenance, maintenance, or operation of **4** to produce or provide food for ~ *vi* **1** to consume food; eat **2** to prey

feeler *n* a tactile appendage (eg a tentacle) of an animal

feeling *n* **1** (a sensation experienced through) the one of the 5 basic physical senses by which stimuli, esp to the skin and mucous membranes, are interpreted by the brain as touch, pressure, and temperature **2** generalized bodily consciousness, sensation, or awareness

feline *adj* of cats or the cat family

female[1] *n* **1** an individual that bears young or produces eggs; *esp* a woman or girl as distinguished from a man or boy **2** a plant or flower with an ovary but no stamens

female[2] *adj* of or being a female

femur *n, pl* **femurs, femora 1** the bone of the hind or lower limb nearest the body; the thighbone ☞ ANATOMY **2** the third segment of an insect's leg counting from the base ☞ INSECT

fen *n* an area of low wet or flooded land

fenestra *n, pl* **fenestrae 1** an oval opening between the middle ear and the vestibule of the inner ear **2** a round opening between the middle ear and the cochlea of the inner ear **3** an opening cut in bone

fenestration *n* **1** an opening in a surface (eg a wall or membrane) **2** the operation of cutting an opening in the bony labyrinth between the inner ear and tympanum as a treatment for deafness

fermentation *n* an enzymatically controlled anaerobic breakdown of an energy-rich compound (eg a carbohydrate to carbon dioxide and alcohol); *broadly* an enzymatically controlled transformation of an organic compound

fern *n* any of a class of flowerless seedless vascular plants; *esp* any of an order resembling flowering plants in having a root, stem, and leaflike fronds but differing in reproducing by spores ☞ EVOLUTION, PLANT

ferruginous, *also* **ferrugineous** *adj* of or containing iron

fertile *adj* **1** (capable of) producing or bearing fruit (in great quantities); productive **2** capable of growing or developing <~ *egg*> **3** capable of breeding or reproducing

fertilization *n* the union of two (haploid) germ cells or gametes to produce a (diploid) zygote which can develop into a new individual

fertil·ize, -ise *vt* to make fertile: eg **a** to inseminate, impregnate, or pollinate **b** to apply a fertilizer to <~ *land*>

fertil·izer, -iser *n* a substance (eg

manure) used to make soil more fertile

fetus *n* a foetus

fever *n* (any of various diseases characterized by) a rise of body temperature above the normal

feverish *also* **feverous** *adj* 1 indicating, relating to, or caused by (a) fever 2 tending to cause or infect with fever

fibre, *NAm chiefly* **fiber** *n* 1 an elongated tapering supportive thick-walled plant cell 2 NERVE 1 3 any of the filaments composing most of the intercellular matrix of connective tissue 4 any of the elongated contractile cells of muscle tissue

fibrescope *n* a flexible instrument using fibre optics for examining inaccessible areas (eg the lining of the stomach)

fibril *n* a small filament or fibre

fibrillation *n* 1 the forming of fibres or fibrils 2 very rapid irregular contractions of muscle fibres (of the heart resulting in a lack of synchronization between heartbeat and pulse)

fibrin *n* a fibrous protein formed from fibrinogen by the action of thrombin, esp in the clotting of blood

fibrinogen *n* a (blood plasma) protein that is produced in the liver and is converted into fibrin during clotting of blood

fibroblast *n* a cell giving rise to connective tissue

fibroid[1] *adj* resembling, forming, or consisting of fibrous tissue

fibroid[2] *n* a benign tumour made up of fibrous and muscular tissue that occurs esp in the uterine wall

fibrosis *n* the abnormal increase of interstitial fibrous tissue in an organ or part of the body

fibrous *adj* 1 containing, consisting of, or resembling fibres 2 characterized by fibrosis

fibula *n, pl* **fibulae, fibulas** the (smaller) outer of the 2 bones of the hind limb of higher vertebrates between the knee and ankle — compare TIBIA ☞ ANATOMY

filament *n* a single thread or a thin flexible threadlike object or part: eg **a** an elongated thin series of attached cells or a very long thin cylindrical single cell (eg of some algae, fungi, or bacteria) **b** the anther-bearing stalk of a stamen ☞ FLOWER

filaria *n, pl* **filariae** any of numerous threadlike nematode worms that usu develop in biting insects and are parasites in the blood or tissues of mammals when adult

filariasis *n, pl* **filariases** infestation with or disease (eg elephantiasis) caused by filarial worms

filial generation *n* a generation in a breeding experiment that is successive to a parental generation

film *n* 1 a thin skin or membranous

covering **2** (dimness of sight resulting from) an abnormal growth on or in the eye

filter feeder *n* an animal (eg a blue whale) adapted to filtering minute organisms or other food from water that passes through its system

filtration *n* passing (as if) through a filter; *also* diffusion <*the kidney produces urine by* ~>

fimbriate, fimbriated *adj* having the edge or extremity bordered by long slender projections; fringed

fin *n* an external membranous part of an aquatic animal (eg a fish or whale) used in propelling or guiding the body

finger *n* any of the 5 parts at the end of the hand or forelimb; *esp* any one other than the thumb

fingerprint *n* the characteristic pattern produced by chromatography or electrophoresis of a particular partially broken down protein or other macromolecule

fire blight *n* a destructive highly infectious disease of apples, pears, and related fruits caused by a bacterium

first-degree burn *n* a mild burn characterized by heat, pain, and reddening of the burned surface but without blistering or charring of tissues — compare SECOND-DEGREE BURN, THIRD-DEGREE BURN

fish *n, pl* **fish, fishes 1** an aquatic

animal — usu in combination <*starfish*> <*cuttlefish*> **2** (the edible flesh of) any of a class of cold-blooded aquatic vertebrates that typicall- have an elongated scaly body, limbs, when present, in the form of fins, and gills

fishery *n* **1** the activity or business of catching fish and other sea animals **2** a place or establishment for catching fish and other sea animals

fish farm *n* a tract of water used for the artificial cultivation of an aquatic life form (eg fishes)

fission *n* reproduction by spontaneous division into 2 or more parts each of which grows into a complete organism

fissiparous *adj* reproducing by fission

fissure *n* a natural cleft between body parts or in the substance of an organ (eg the brain)

fistula *n, pl* **fistulas, fistulae** an abnormal or surgically made passage leading from an abscess or hollow organ to the body surface or between hollow organs

fit *adj* **-tt- 1** adapted to the environment so as to be capable of surviving **2** HEALTHY 1

fix *vt* **1** to change into a stable compound or available form <*bacteria that* ~ *nitrogen*> **2** to kill, harden, and preserve for microscopic study

fixation *n* **1** an (obsessive or unhealthy) attachment or preoccupation

2 a concentration of the libido on infantile forms of gratification <~ at the oral stage>

flaccid adj limp — compare TURGID

flagellate¹, flagellated adj 1 having flagella 2 shaped like a flagellum

flagellate² n a protozoan or algal cell that has a flagellum

flagellum n, pl **flagella** also **flagellums** any of various elongated filament-shaped appendages of plants or animals; esp one that projects singly or in groups and powers the motion of a microorganism

flame cell n a hollow cell that has a tuft of cilia and is part of the excretory system of various lower invertebrates

flatfish n any of an order of marine fishes (eg the flounders and soles) that swim on one side of the flattened body and have both eyes on the upper side

flatus n gas generated in the stomach or intestines

flatworm n a platyhelminth

flavin n any of several yellow pigments occurring as part of the coenzymes of flavoproteins

flavine n acriflavine or a similar yellow dye used as an antiseptic

flavoprotein n an enzyme that contains a flavin and often a metal and plays a major role in biological oxidation reactions

flea n any of an order of wingless bloodsucking jumping insects that feed on warm-blooded animals

flesh n 1 the soft, esp muscular, parts of the body of a (vertebrate) animal as distinguished from visceral structures, bone, hide, etc 2 a fleshy (edible) part of a plant or fruit

fleshy adj 1 consisting of or resembling flesh 2 marked by (abundant) flesh; esp corpulent 3 succulent, pulpy

flex vt 1 to bend (a limb or joint) 2 to move (a muscle or muscles) so as to flex a limb or joint

flexor n a muscle that flexes a joint of a limb

flipper n a broad flat limb (eg of a seal) adapted for swimming

float n a sac containing air or gas and buoying up the body of a plant or animal

floating adj located out of the normal position <a ~ kidney>

floating rib n a rib (eg any of the last 2 pairs in human beings) that has no attachment to the sternum — compare FALSE RIB

flocculate vb to (cause to) form a flocculent mass of small loosely aggregated bits of material suspended in or precipitated from a liquid <~ clay>

flock n sing or pl in constr a group of birds or mammals assembled or herded together

floor n the lower inside surface of a hollow structure (eg a bodily part)

flora n, pl **floras** also **florae** 1 a treatise on, or a work used to identify, the plants of a region **2** plant life (of a region, period, or special environment) — compare FAUNA

floral *adj* of flowers or a flora

floral leaf *n* a modified leaf (eg a sepal or petal) occurring as part of the inflorescence of a plant

floret *n* any of the small flowers forming the head of a (composite) plant

flower¹ *n* **1** a blossom, inflorescence **2** a shoot of a higher plant bearing leaves modified for reproduction to form petals, sepals, ovaries, and anthers ⊙

flower² *vi* to produce flowers; blossom

flowering plant *n* a plant that produces flowers, fruit, and seed; an angiosperm ⟹ EVOLUTION, PLANT

fluid *n* a liquid in the body of an animal or plant <*cerebrospinal* ~>

fluke¹ *n* **1** a flatfish **2** a liver fluke or related trematode worm

fluke² *n* either of the lobes of a whale's tail

flush *n* **1** a sudden increase, esp of new plant growth **2** a transitory sensation of extreme heat; *specif* HOT FLUSH

flutter *vi* **1** to flap the wings rapidly **2** to beat or vibrate in irregular spasms <*his pulse* ~ed>~ *vt* to cause to flutter

focus¹ *n, pl* **focuses, foci 1** a point at which rays (of light, heat, or

sound) converge or from which they (appear to) diverge **2** an adjustment for clear vision **3** a localized area of disease or the chief site of a generalized disease

focus² *vb* **focused, focussed** *vt* to bring into focus ~ *vi* to adjust one's eye to a particular range

foetus *n* an unborn or unhatched vertebrate; *specif* a developing human from usu 3 months after conception to birth ⟹ REPRODUCTION

foliaceous *adj* **1** of or resembling a foliage leaf **2** consisting of thin plates

foliage leaf *n* an ordinary green leaf as distinguished from a floral leaf, scale, or bract

folic acid *n* a vitamin of the vitamin B complex that is found esp in green leafy vegetables and liver and whose lack in the diet results in anaemia

follicle *n* **1a** a small anatomical cavity or deep narrow depression **b** GRAAFIAN FOLLICLE **2** a dry 1-celled many-seeded fruit that has a single carpel and opens along 1 line only

follicle-stimulating hormone *n* a hormone produced by the front lobe of the pituitary gland that stimulates the growth of the ovum-containing Graafian follicles in women and activates sperm-forming cells in men

food *n* **1** (minerals, vitamins, etc

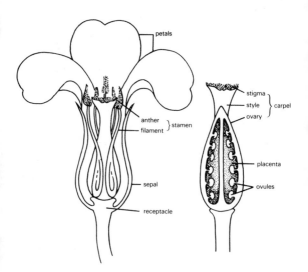

Structure and function of a typical flower

A flower is a specialized reproductive shoot producing seeds which give rise to the next generation. The flower parts are attached to the receptacle in rings, or whorls. The innermost whorl of one or more carpels is encircled by a whorl of stamens. In most flowers, these reproductive organs are surrounded by a whorl of petals. Coloured, scented petals with a nectary at the base attract insects which, as they collect nectar, bring about cross-pollination by transferring pollen from the anthers of one flower to the stigma of another. An outer whorl of leaflike sepals encloses and protects the other flower parts as they develop at the bud stage.

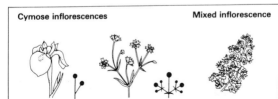

Racemose inflorescences

capitulum (dandelion)

corymb (yarrow)

panicle (wild oat)

raceme (foxglove)

spike (greater plantain)

compound umbel (wild carrot)

Cymose inflorescences

Mixed inflorescence

simple (yellow iris)

compound (greater stitchwort)

thyrsus (horse chestnut)

together with) material consisting essentially of protein, carbohydrate, and fat taken into the body of a living organism and used to provide energy and sustain processes (eg growth and repair) essential for life **2** inorganic substances absorbed (eg in gaseous form or in solution) by plants

food chain n a hierarchical arrangement of organisms ordered according to each organism's use of the next as a food source

food vacuole n a vacuole (eg in an amoeba) in which ingested food is digested

food web n all the interacting food chains in an ecological community
☞ ECOLOGY

foot n, pl **feet 1** the end part of the vertebrate leg on which an animal stands **2** an organ of locomotion or attachment of an invertebrate animal, esp a mollusc

foot-and-mouth,   foot-and-mouth disease n a contagious virus disease, esp of cloven-footed animals, marked by small ulcers in the mouth, about the hoofs, and on the udder and teats

forage n food for animals, esp when taken by browsing or grazing

foraminifer, foraminiferan n, pl **foraminifera, foraminifers, foraminiferans** any of an order of chiefly marine amoeba-like single-celled animals usu having hard perforated calcium-containing shells that form the bulk of chalk

forearm n (the part in other vertebrates corresponding to) the human arm between the elbow and the wrist

forebrain n (the telencephalon and other parts of the adult brain that develop from) the front of the 3 primary divisions of the embryonic vertebrate brain

foreign adj occurring in an abnormal situation in the living body and commonly introduced from outside

foreleg n a front leg, esp of ˄a quadruped

forelimb n an arm, fin, wing, or leg that is (homologous to) a foreleg

foreskin n a fold of skin that covers the glans of the penis

form n a body (eg of a person), esp in its external appearance or as distinguished from the face

formative adj capable of alteration by growth and development <~ tissues>

formicary n an ant nest

fossa n, pl **fossae** an anatomical pit or depression

fossil[1] n a relic of an animal or plant of a past geological age, preserved in the earth's crust

fossil[2] adj **1** extracted from the earth and derived from the remains of living things <coal is a ~ fuel> **2** preserved in a mineralized or petrified form from a past geological age

fossil·ize, -ise *vt* to convert into a fossil ~ *vi* to become fossilized

fossorial *adj* adapted to digging

fovea *n, pl* **foveae** a small anatomical pit; *esp* FOVEA CENTRALIS

fovea centralis *n, pl* **foveae centrales** an area of the retina with numerous cones but no rods where vision is acute

fowl pest *n* a fatal infectious virus disease of domestic poultry

fraternal *adj, of twins* derived from 2 ova

free *adj* not (permanently) united with, attached to, or combined with sthg else; separate <~ *oxygen*>

free association *n* the expression of conscious thoughts, ideas, etc used esp in psychoanalysis to reveal unconscious processes; *esp* (the reporting of) the first thought, image, etc that comes to mind in response to a given stimulus (eg a word)

free-living *adj, of a living organism* neither parasitic nor symbiotic

free-swimming *adj, of an animal that lives in water* able to swim about; not attached to a rock or other object

frequency *n* the number or proportion of individuals in a single class when objects are classified according to variations in a set of attributes

frequency distribution *n* DISTRIBUTION 3

freshwater *adj* of or living in fresh water

Freudian *adj* of or conforming to the psychoanalytic theories or practices of S Freud

Freudian slip *n* a slip of the tongue that is held to reveal some unconscious aspect of the speaker's mind

friable *adj* easily crumbled

frog *n* **1** any of a suborder of tailless smooth-skinned web-footed largely aquatic leaping amphibians **2** the triangular horny pad in the middle of the sole of a horse's foot

frogspawn *n* (a gelatinous mass of) frog's eggs

frontal lobe *n* the front lobe of either cerebral hemisphere

fructification *n* forming or producing fruit

fructose *n* a (very sweet) disaccharide sugar that occurs esp in fruit juices and honey

frugivorous *adj* feeding on fruit

fruit[1] *n* **1** the (edible) reproductive body of a flowering plant; *esp* one having a sweet pulp associated with the seed **2** the ripened fertilized ovary of a flowering plant together with its contents **3** offspring, progeny

fruit[2] *vb* to (cause to) bear fruit

fruiting body *n* a plant organ specialized for producing spores

fry *n, pl* **fry 1** recently hatched or very small (adult) fishes **2** the young of other animals, esp when occurring in large numbers

full blood *n* (an individual having) descent from parents both of the same pure breed

full-blown *adj* at the height of bloom

full-term *adj* born after a pregnancy of normal length — compare PRE-MATURE

function *n* any of a group of related actions contributing to a larger action

functional *adj* 1 affecting physiological or psychological functions but not organic structure <~ *heart disease*> — compare ORGANIC 1b 2 (capable of) performing a function

fungicide *n* a substance used for destroying or preventing the growth of a fungus

fungoid *adj* resembling, characteristic of, or being a fungus

fungous *adj* of, like, or caused by a fungus or fungi

fungus *n, pl* **fungi** *also* **funguses** any of a class of parasitic or saprophytic lower plants lacking chlorophyll and including moulds, rusts, mildews, smuts, mushrooms, and toadstools ☞ EVOLUTION, PLANT

fusiform *adj* tapering towards each end <~ *bacteria*>

galactose *n* a monosaccharide sugar that is less soluble and less sweet than glucose

galea *n* an anatomical part suggesting a helmet (eg the upper lip of the corolla of a mint; or the outer lobe of the maxilla in certain insects)

gall *n* 1 bile 2 a skin sore caused by rubbing 3 a diseased swelling of plant tissue produced by infection with fungi, insect parasites, etc

gall bladder *n* a membranous muscular sac in which bile from the liver is stored

gallinaceous *adj* of an order of (ground-living) birds including the pheasants, turkeys, grouse, and the common domestic fowl

gallstone *n* a calculus formed in the gall bladder or bile ducts

gametangium *n, pl* **gametangia** a (plant) organ in which gametes are developed

gamete *n* a mature germ cell with a single set of chromosomes capable of fusing with another gamete of the other sex to form a zygote from which a new organism develops

gametophyte *n* (a member of) the generation that bears sex organs, of a plant with alternation of generations — compare SPOROPHYTE

gamma globulin *n* any of several immunoglobulins in blood or serum including most antibodies

gamopetalous *adj, of a flower* having the corolla composed of united petals

gamosepalous *adj, of a flower* having the calyx composed of united sepals

ganglion *n, pl* **ganglia** *also* **ganglions** 1 a small cyst on a joint membrane or tendon sheath 2 a mass of nerve cells outside the brain or spinal cord; *also* NUCLEUS **b**

gangrene *n* local death of the body's soft tissues due to loss of blood supply

gas gangrene *n* often rapidly progressive gangrene marked by impregnation of the (dying) tissue with gas and caused by infection with a clostridial bacterium

gastric *adj* of the stomach

gastric juice *n* a thin acidic digestive liquid secreted by glands in the lining of the stomach

gastrin *n* a polypeptide hormone secreted by the stomach lining that induces secretion of gastric juice

gastroenteritis *n* inflammation of the lining of the stomach and the intestines, usu causing painful diarrhoea

gastropod *n* any of a large class of molluscs (eg snails) usu with a distinct head bearing sensory organs

gastrula *n*, *pl* **gastrulas, gastrulae** the embryo of a metazoan, animal at the stage in its development succeeding the blastula stage and consisting of a hollow 2-layered cellular cup— compare BLASTULA, MORULA

Gaussian distribution *n* NORMAL DISTRIBUTION

gel *n* a colloid in a more solid form than a sol

gelatin, gelatine *n* a glutinous material obtained from animal tissues by boiling; *esp* a protein used esp in food (eg to set jellies) and photography

gelatinous *adj* resembling gelatin or jelly, esp in consistency; viscous

geminate *vb* to make or become paired or doubled

gemma *n*, *pl* **gemmae** a bud; *broadly* an asexual plant reproductive body

gene *n* a unit of inheritance that is carried on a chromosome, controls transmission of hereditary characters, and consists of DNA or, in some viruses, RNA

genera *pl of* GENUS

general·ization, -isation *n* the occurring of a response to a stimulus similar but not identical to a reference stimulus

generation *n* **1** *sing or pl in constr* **a** a group of living organisms constituting a single step in the line of descent from an ancestor **b** a group of individuals born and living at the same time **2** the average time between the birth of parents and that of their offspring **3** the producing of offspring; procreation

generative *adj* having the power or function of generating, originating, producing, reproducing, etc

generic *adj* (having the rank) of a biological genus

genetic *adj* of or involving genetics

genetic code *n* the sequence of bases in DNA or RNA strands that

gen

forms the biochemical basis of heredity and determines the specific amino acid sequence in proteins

genetics *n pl but sing in constr* 1 the biology of (the mechanisms and structures involved in) the heredity and variation of organisms 2 the genetic make-up of an organism, type, group, or condition

genital *adj* 1 of or being the genitalia or another sexual organ 2 of or characterized by the final stage of sexual development in which oral and anal impulses are replaced by gratification obtained from (sexual) relationships — compare ANAL, ORAL

genitalia, genitals *n pl* the (external) reproductive and sexual organs

genitourinary *adj* of the (functions of the) genital and urinary organs

genome *n* a single set of an organism's chromosomes with the genes they contain

genotype *n* the genetic constitution of an individual or group — compare PHENOTYPE

genus *n, pl* **genera** a category in the classification of living things ranking between the family and the species

geotaxis *n* a response of a cell or organism to the force of gravity

geotropism *n* a tropism (eg in the downward growth of roots) in which gravity is the orienting factor

germ *n* 1 a small mass of cells

capable of developing into (a part of) an organism 2 the embryo of a cereal grain that is usu separated from the starchy endosperm during milling 3 a (disease-causing) microorganism

German measles *n pl but sing or pl in constr* a virus disease that is milder than typical measles but is damaging to the foetus when occurring early in pregnancy

germ cell *n* (a cell from which is derived) an egg or sperm cell

germicide *n* sthg that kills germs

germinate *vt* to cause to sprout or develop ~ *vi* to begin to grow; sprout

germ plasm *n* the hereditary material of the germ cells; the genes

gestalt *n, pl* **gestalten, gestalts** a structure, pattern, etc (eg a melody) that as an object of perception constitutes a functional unit with properties not derivable from the sum of its parts

Gestalt psychology *n* the study of perception and behaviour using the theory that perceptions, reactions, etc are gestalts

gestation *n* the carrying of young in the uterus; pregnancy

gibberellin *n* any of several plant hormones that promote shoot growth

gill *n* 1 an organ, esp of a fish, for oxygenating blood using the oxygen dissolved in water 2 the flesh under or about the chin or jaws —

usu pl with sing meaning **3** any of the radiating plates forming the undersurface of the cap of some fungi (eg mushrooms)

gill cover n the operculum

gingiva n, pl **gingivae** [1]GUM

girdle[1] n **1** a bony ring at the front and rear end of the trunk of vertebrates supporting the arms or legs <pelvic ~> **2** a ring made by the removal of the bark and cambium round a plant stem or tree trunk

girdle[2] vt **girdling** to cut a girdle round (esp a tree), usu in order to kill

gizzard n **1** a muscular enlargement of the alimentary canal of birds that immediately follows the crop and has a tough horny lining for grinding food **2** a thickened part of the alimentary canal of some animals (eg an earthworm) similar in function to the crop of a bird

glabrous adj smooth; esp having a surface without hairs or projections

gland n **1** (an animal structure that does not secrete but resembles) an organ that selectively removes materials from the blood, alters them, and secretes them esp for further use in the body or for elimination **2** any of various secreting organs (eg a nectary) of plants

glandular adj of, involving, or being (the cells or products of) glands

glandular fever n INFECTIOUS MONONUCLEOSIS

glans n, pl **glandes** a conical vascular part at the end of the penis or clitoris

glaucoma n increased pressure within the eyeball (leading to damage to the retina and gradual loss of vision)

glaucous adj **1** pale yellowy green **2** esp of plants or plant parts of a dull blue or bluish-green colour **3** of a plant or fruit having a powdery or waxy coating giving a frosted appearance

gley n a sticky clay formed under the surface of some waterlogged soils

globin n a colourless protein obtained by removal of haem from esp haemoglobin

globular adj globe- or globule-shaped <~ proteins>

globulin n any of a class of widely occurring proteins that are soluble in dilute salt solutions

glomerule n a compact clustered flower head like that of a composite plant

glomerulus n, pl **glomeruli** a small coiled or intertwined mass; specif the compact mass of capillaries at the end of each nephron of the kidneys of vertebrates

glossa n, pl **glossae** also **glossas** a (structure like a) tongue, esp in (the labium of) an insect

glottis n, pl **glottises, glottides** (the

glu

structures surrounding) the elongated space between the vocal cords — compare EPIGLOTTIS

glucagon *n* a hormone produced esp by the pancreatic islets of Langerhans that promotes an increase in the sugar content of the blood by increasing the rate of breakdown of glycogen in the liver

glucocorticoid *n* any of several corticosteroids (eg cortisol) that affect metabolic processes and are used in medicine (eg in treating rheumatoid arthritis) because they suppress inflammation and inhibit the activity of the immune system

glucose *n* a sweet (dextrorotatory form of a) monosaccharide sugar that occurs widely in nature and is the usual form in which carbohydrate is assimilated by animals

glucoside *n* a glycoside (that yields glucose on hydrolysis)

glume *n* a chaffy bract, specif in the spikelet of grasses

glutamic acid *n* an acidic amino acid found in most proteins

glutamine *n* an amino acid that is a chemical base and is found in nearly all proteins

gluten *n* an elastic protein substance, esp of wheat flour, that gives cohesiveness to dough

glyceraldehyde *n* a sweet compound formed as an intermediate in carbohydrate metabolism

glycerol *n* a trihydroxy alcohol usu

obtained by the saponification of fats

glycine *n* a sweet amino acid found in most proteins

glycogen *n* a polysaccharide that is the chief storage carbohydrate of animals

glycogenesis *n* the formation of (sugar from) glycogen

glycolysis *n* the enzymatic breakdown of a carbohydrate with the production of energy for storage in the cell

glycoside *n* any of numerous sugar derivatives in which a nonsugar group is attached by an oxygen or nitrogen atom and that on hydrolysis yield a sugar

glycosuria *n* the presence of abnormal amounts of sugar in the urine

goblet cell *n* a mucus-secreting epithelial cell shaped like a goblet and found in mucous membranes (eg of the intestines)

goitre, *NAm chiefly* **goiter** *n* an abnormal enlargement of the thyroid gland visible as a swelling of the front of the neck

Golgi *adj* of the Golgi apparatus or bodies <~ *vesicles*>

Golgi apparatus *n* a cytoplasmic organelle that appears in electron microscopy as a series of parallel (vesicular) membranes and is concerned with secretion of cell products

Golgi body *n* (a discrete particle of) the Golgi apparatus

gonad *n* any of the primary sex glands (eg the ovaries or testes)

gonadotrophic, gonadotropic *adj* acting on or stimulating the gonads

gonadotrophin, gonadotropin *n* a gonadotrophic hormone (eg follicle-stimulating hormone)

gonidium *n, pl* **gonidia** an asexual reproductive cell or group of cells in or on a gametophyte

gonococcus *n, pl* **gonococci** the pus-producing bacterium that causes gonorrhoea

gonorrhoea, *chiefly NAm* **gonorrhea** *n* a venereal disease in which there is inflammation of the mucous membranes of the genital tracts caused by gonococcal bacteria

gorgonian *n* any of an order of colonial anthozoan polyps

gout *n* painful inflammation of the joints, esp that of the big toe, resulting from a metabolic disorder in which there is an excessive amount of uric acid in the blood

Graafian follicle *n* a vesicle in the ovary of a mammal enclosing a developing egg

graft¹ *vt* **1** to cause (a plant scion) to unite with a stock; *also* to unite (plants or scion and stock) to form a graft **2** to propagate (a plant) by grafting **3** to implant (living tissue) surgically ~ *vi* **1** to become grafted **2** to perform grafting

graft² *n* **1a** a grafted plant **b** (the point of insertion upon a stock of)

a scion **2** (living tissue used in) grafting

grain *n* a seed or fruit of a cereal grass; *also* (the seeds or fruits collectively of) the cereal grasses or similar food plants

gramineous, graminaceous *adj* of a grass

gram-negative *adj, of bacteria* not holding the purple dye when stained by Gram's method

gram-positive *adj of bacteria* holding the purple dye when stained by Gram's method

Gram's method *n* the treatment of bacteria with a solution of iodine and potassium iodide after staining with gentian violet so that some species are decolorized and some remain coloured

granular *adj* (apparently) consisting of granules; having a grainy texture

granule *n* a small particle

granulocyte *n* any of various white blood cells that have cytoplasm containing large numbers of conspicuous stainable granules and a nucleus with many lobes — compare AGRANULOCYTE, BASOPHIL, EOSINOPHIL

granum *n, pl* **grana** any of the stacks of thin layers of chlorophyll-containing material in plant chloroplasts

grass *n* **1** any of a large family of plants with slender leaves and (green) flowers in small spikes or

gra

clusters, that includes bamboo, wheat, rye, corn, etc **2** grass leaves or plants

gravel *n* (a stratum or surface of) loose rounded fragments of rock mixed with sand

graze[1] *vi* to feed on growing herbage ~ *vt* **1** to crop and eat (growing herbage) **2** to feed on the herbage of (eg a pasture)

graze[2] *vt* to abrade, scratch <~d *her elbow*>

great *adj* of a relatively large kind — in plant and animal names

green alga *n* any of a class of algae with well-defined chloroplasts in which the chlorophyll is not masked by other pigments

greenfly *n, pl* **greenflies**, *esp collectively* **greenfly** *Br* (an infestation by) any of various green aphids that are destructive to plants

green manure *n* a herbaceous crop (eg clover) ploughed under while green to enrich the soil

gregarious *adj* **1** of or living in a social group **2** *of a plant* growing in a cluster or colony

grey matter *n* brownish-grey nerve tissue, esp in the brain and spinal cord, containing nerve-cell bodies as well as nerve fibres

ground cover *n* (all the) low-growing plants (in a forest except young trees)

group therapy *n* the treatment of several individuals (with similar psychological problems) simultaneously through group discussion and mutual aid

grow *vi* **grew, grown 1** to spring up and develop to maturity (in a specified place or situation) **2** to assume some relation (as if) through a process of natural growth <*2 tree trunks* grown *together*> **3** to increase in size by addition of material (eg by assimilation into a living organism or by crystallization)

growth *n* **1a** (a stage in the process of) growing **b** progressive development **2a** sthg that grows or has grown **b** a tumour or other abnormal growth of tissue

growth factor *n* a substance (eg a vitamin) necessary for the growth of an organism

growth hormone *n* **1** a polypeptide growth-regulating hormone of vertebrates that is secreted by the front lobe of the pituitary gland **2** any of various plant substances (eg an auxin or gibberellin) that promote growth

guanine *n* a purine base that is one of the 4 bases whose order in a DNA or RNA chain codes genetic information — compare ADENINE, CYTOSINE, THYMINE, URACIL

guanosine *n* a nucleoside containing guanine

guard cell *n* either of the 2 crescent-shaped cells that border and open and close a plant stoma

gullet *n* the oesophagus; *broadly* the throat

gum[1] *n* (the tissue that surrounds the teeth and covers) the parts of the jaws from which the teeth grow ⏵ ANATOMY

gum[2] *n* **1** any of numerous polysaccharide plant substances that are gelatinous when moist but harden on drying — compare MUCILAGE **2** any of various substances (eg a mucilage or gum resin) that exude from plants

gut *n* **1** (a part of) the alimentary canal **2** the belly or abdomen

guttural *adj* of the throat

gymnosperm *n* any of a class of woody vascular seed plants (eg conifers) that produce naked seeds not enclosed in an ovary — compare ANGIOSPERM ⏵ PLANT

gynaecology *n* a branch of medicine that deals with diseases and disorders (of the reproductive system) of women

gynandromorph *n* an (abnormal) individual having characters of both sexes in different parts of the body

gynandrous *adj, of a flower, esp an orchid* having the male and female parts united in a column

gynoecium *n, pl* **gynoecia** all the female parts of a flower

habit *n* **1** an acquired pattern or mode of behaviour **2** addiction <*a drug* ~> **3** characteristic mode of growth, occurrence, or appearance (eg of a plant or crystal)

habitat *n* the (type of) place where a plant or animal naturally grows or lives

haem, *chiefly NAm* **heme** *n* a deep red iron-containing compound that occurs esp as the oxygen-carrying part of haemoglobin

haemagglutinate *vt* to cause agglutination of (red blood cells)

haemal *adj* **1** of the blood (vessels) **2** of or situated on the same side of the spinal cord as that on which the heart is placed

haematology *n* the biology and medicine of (diseases of) the blood and blood-forming organs

haematoma *n* a tumour or swelling containing blood; BRUISE 1

haemocoele, haemocoel *n* a body cavity in arthropods or some other invertebrates that normally contains blood and functions as part of the circulatory system

haemocyanin *n* a colourless copper-containing respiratory pigment found in the blood of various arthropods and molluscs that is analogous to the haemoglobin of higher animals

haemocyte *n* a blood cell, esp of an invertebrate animal

haemocytometer *n* an instrument for counting (blood) cells suspended in a liquid, usu when viewed under a microscope

haemodialysis *n* purification of the

blood (of sby whose kidneys have failed) by dialysis

haemoglobin *n* an iron-containing protein that occurs in the red blood cells of vertebrates and is the means of oxygen transport from the lungs to the body tissues

haemolymph *n* a circulatory fluid of various invertebrate animals that is functionally comparable to the blood and lymph of vertebrates

haemolysis *n* dissolution of red blood cells with release of haemoglobin

haemophilia *n* delayed clotting of the blood with consequent difficulty in controlling bleeding even after minor injuries, occurring as a hereditary defect, usu in males

haemophiliac *n or adj* (sby) suffering from haemophilia

haemopoiesis *n* the formation of blood cells in the bone marrow and lymphoid tissue

haemorrhage *n* a (copious) loss of blood from the blood vessels

haemorrhoid *n* a mass of dilated veins in swollen tissue round or near the anus — usu pl with sing. meaning

haemostasis *n* arrest of bleeding

hair *n* 1 (a structure resembling) a slender threadlike outgrowth on the surface of an animal; *esp* (any of) the many usu pigmented hairs that form the characteristic coat of a mammal 2 the coating of hairs,

esp on the human head or other body part

half-bred *adj* having 1 purebred parent

half-breed *n* the offspring of parents of different races

half-life *n* the time required for half of **a** the atoms of a radioactive substance to become disintegrated **b** a drug or other substance to be eliminated from an organism by natural processes

halitosis *n* (a condition of having) offensively smelling breath

hallucination *n* 1 the perception of sthg apparently real to the perceiver but which has no objective reality, *also* the image, object, etc perceived 2 a completely unfounded or mistaken impression or belief

hallucinatory *adj* 1 tending to produce hallucination <~ *drugs*> 2 resembling or being a hallucination

hallucinogen *n* a substance (eg LSD) that induces hallucinations

haltere *also* **halter** *n, pl* **halteres** either of a pair of club-shaped sensory organs in a two-winged fly that maintain equilibrium in flight

hammer *n* the malleus

hamstring *n* 1 either of 2 groups of tendons at the back of the human knee 2 a large tendon above and behind the hock of a quadruped

hand *n* 1 (the segment of the forelimb of vertebrate animals corresponding to) the end of the forelimb of human beings, monkeys, etc when modified as a grasping organ **2** a part (eg the chela of a crustacean) serving the function of or resembling a hand

haploid *adj* having half the number of chromosomes characteristic of somatic cells <*gametes are usually* ~> — compare DIPLOID, POLYPLOID

hapten *n* a small (separable) part of an antigen that reacts specifically with an antibody

haptic, haptical *adj* relating to or based on the sense of touch

harden *vt* to inure (eg plants) to cold or other unfavourable environmental conditions — often + *off*

hardening *n* sclerosis <~ *of the arteries*>

hard palate *n* the bony front part of the palate forming the roof of the mouth **æ** SENSE ORGAN

hardpan *n* a hard compact soil layer

hardwood *n* **1** (the wood of) a broad-leaved as distinguished from a coniferous tree **2** mature woody tissue — compare SAPWOOD

hardy *adj* capable of withstanding adverse conditions; *esp* capable of living outdoors over winter without artificial protection <~ *plants*>

harvest *vt* **1** to gather in (a crop); reap **2** to gather (a natural product) as if by harvesting <~ *bacteria*> ~ *vi* to gather in a food crop

hastate *adj* shaped like the (trian-gular head of) a spear <*a ~ leaf*> **ᴈ** PLANT

hatch *vi* **1** to emerge from an egg or pupa **2** to incubate eggs; brood **3** to give forth young <*the egg* ~ed> ~ *vt* to produce (young) from an egg by applying heat

haulm *n* **1** the stems or tops of potatoes, peas, beans, etc (after the crop has been gathered) **2** *Br* an individual plant stem

haustellum *n, pl* **haustella** a mouth part (eg of an insect) adapted to suck blood, plant juices, etc

hay *n* herbage, esp grass, mowed and cured for fodder

head *n, pl* **heads 1** the upper or foremost division of the body containing the brain, the chief sense organs, and the mouth **2** a capitulum **3** the foliaged part of a plant, esp when consisting of a compact mass of leaves or fruits **4** the part of a boil, pimple, etc at which it is likely to break

hear *vb* **heard** *vt* to perceive (sound) with the ear **~** *vi* to have the capacity of perceiving sound

hearing *n* the one of the 5 basic physical senses by which waves received by the ear are interpreted by the brain as sounds varying in pitch, intensity, and timbre

heart *n* a hollow muscular organ that by its rhythmic contraction acts as a force pump maintaining the circulation of the blood **ᴈ** RESPIRATION

heart attack n an instance of abnormal functioning of the heart; esp CORONARY THROMBOSIS

heartbeat n a single complete pulsation of the heart

heart block n incoordination of the beating of the atria and ventricles of the heart resulting in a decreased output of blood

heartburn n a burning pain behind the lower part of the breastbone usu resulting from spasm of the stomach or throat muscles

heart failure n (inability of the heart to perform adequately often leading to) cessation of the heartbeat and death

heart-lung machine n a mechanical pump that shunts the body's blood away from the heart and maintains the circulation and respiration during heart surgery

heat n readiness for sexual intercourse in a female mammal; specif oestrus — usu in on heat or (chiefly NAm) in heat

heath n 1 a tract of wasteland 2 a large area of level uncultivated land usu with poor peaty soil and bad drainage

heatstroke n overheating of the body resulting from prolonged exposure to high temperature and leading to (fatal) collapse

heel n 1 (the back part of the hind limb of a vertebrate corresponding to) the back of the human foot below the ankle and behind the arch or an anatomical structure resembling this 2 the base of a tuber or cutting of a plant used for propagation

HeLa adj of, derived from, or being a particular strain of human cells kept continuously in tissue culture

heliotaxis n the response of a cell or organism to the stimulus of sunlight

heliotropism n a tropism in which sunlight is the orienting stimulus, esp in flowers

helix n, pl **helices** also **helixes** 1 sthg spiral in form < DNA ~ > 2 the rim curved inwards of the external ear ☞ SENSE ORGAN

helminth n an (intestinal) worm — used technically

helminthology n a branch of zoology concerned with helminths; esp the study of parasitic worms

heme n, chiefly NAm haem

hemicellulose n any of various polysaccharides of plant cell walls that are less complex than cellulose

hemichordate n any of a division of marine chordate animals with an outgrowth of the pharyngeal wall prob homologous with the notochord of higher chordates

hemiplegia n paralysis of (part of) 1 lateral half of the body

heparin n a polysaccharide that is found esp in liver and is injected to slow the clotting of blood, esp in the treatment of thrombosis

hepatic *adj* of or resembling the liver

hepatitis *n, pl* **hepatitides** (a condition marked by) inflammation of the liver

herb *n* **1** a seed plant that does not develop permanent woody tissue and dies down at the end of a growing season **2** a plant (part) valued for its medicinal, savoury, or aromatic qualities <*cultivated a ~ garden*>

herbaceous *adj* of, being, or having the characteristics of a (part of a) herb

herbage *n* (the succulent parts of) herbaceous plants (eg grass), esp when used for grazing

herbal *n* a book about (the medicinal properties of) plants

herbarium *n, pl* **herbaria** (a place containing) a collection of dried plant specimens usu mounted and systematically arranged for reference

herbicide *n* sthg used to destroy or inhibit plant growth

herbivore *n* a plant-eating animal

hereditary *adj* **1** genetically transmitted or transmissible from parent to offspring **2** of inheritance or heredity

heredity *n* **1** the sum of the qualities and potentialities genetically derived from one's ancestors **2** the transmission of qualities from ancestor to descendant through a

mechanism lying primarily in the chromosomes

heritable *adj* capable of being genetically transmitted

hermaphrodite *n* an animal or plant having both male and female reproductive organs

hernia *n, pl* **hernias, herniae** a protrusion of (part of) an organ through a wall of its enclosing cavity (eg the abdomen)

herpes simplex *n* a virus disease marked by groups of watery blisters on the skin or mucous membranes (eg of the mouth, lips, or genitals)

herpes zoster *n* shingles

herpetology *n* zoology dealing with reptiles and amphibians

hesperidium *n, pl* **hesperidia** an orange or similar fruit with a leathery rind and a pulp divided into sections

heterochromatic *adj* of heterochromatin

heterochromatin *n* densely staining chromatin that appears as nodules in or along chromosomes and contains relatively few genes

heteroecious, *chiefly NAm* **heterecious** *adj* passing different stages in the life cycle on alternative and often unrelated hosts

heterogamete *n* either of a pair of gametes that differ in form, size, or behaviour and occur typically as large nonmotile oogametes and small motile sperms

het

heterogametic *adj* forming 2 kinds of germ cells of which one produces male offspring and the other female offspring <the human male is ~>

heterogamous *adj* **1** bearing flowers of two kinds **2** having or characterized by fusion of unlike gametes — compare ANISOGAMOUS, ISOGAMOUS

heterogamy *n* (the condition of having) sexual reproduction involving fusion of unlike gametes

heterogenesis *n* ALTERNATION OF GENERATIONS

heterograft *n* a graft of tissue taken from a donor of one species and grafted into a recipient of another species

heterokaryon *also* **heterocaryon** *n* a cell in the mycelium of a fungus that contains 2 or more genetically unlike nuclei

heteromorphic, heteromorphous *adj* exhibiting diversity of form or forms <~ *pairs of chromosomes*>

heterophyte *n* a plant (eg a saprophyte or parasite) that is dependent for food materials upon other organisms or their products

heterosexual *adj or n* (of or being) sby having a sexual preference for members of the opposite sex — compare HOMOSEXUAL

heterosis *n* a marked vigour or capacity for growth often shown by crossbred animals or plants

heterospory *n* **1** the production of sexual spores of more than one kind **2** the production of microspores and megaspores as in ferns and seed plants

heterothallic *adj* having 2 or more genetically incompatible haploid phases that function as separate sexes — used esp of certain algae and fungi

heterotrophic *adj* needing ready-made complex organic compounds for essential metabolic processes — compare AUTOTROPHIC

heterozygote *n* an animal, plant, or cell having dissimilar alleles (eg 1 dominant and 1 recessive) of a particular gene — compare HOMOZYGOTE

hexapod *n or adj* (an insect) having 6 feet

hexose *n* a monosaccharide (eg glucose) containing 6 carbon atoms in the molecule

hiatus *n* an (abnormal) anatomical gap or passage

hibernate *vi* to pass the winter in a torpid or resting state — compare AESTIVATE

hilum *n, pl* **hila** **1** a scar on a seed (eg a bean) marking the point of attachment of the ovule to its stalk **2** the nucleus of a starch grain **3** a notch, opening, etc in a bodily part, usu where a vessel, nerve, etc enters

hindbrain *n* (the cerebellum, pons, and other parts of the adult brain that develop from) the rear of the 3

102

primary divisions of the embryonic vertebrate brain

hinge joint *n* a joint between bones (eg at the elbow) that permits movement in 1 plane only

hip *n* **1** the projecting region at each side of the lower or rear part of the mammalian trunk formed by the pelvis and upper part of the thigh **2** HIP JOINT

hipbone *n* INNOMINATE BONE

hip joint *n* the joint between the femur and the hipbone

hippocampus *n, pl* **hippocampi** a curved elongated ridge of nervous tissue inside each hemisphere of the brain

hircine *adj* goatlike

hirsute *adj* covered with (coarse stiff) hairs

hispid *adj* covered with bristles, stiff hairs, etc <*a ~ plant*> — compare PUBESCENT 2

histamine *n* an amine that is a neurotransmitter in the autonomic nervous system and whose release under certain conditions causes an allergic reaction

histidine *n* an amino acid that is a chemical base and is found in most proteins

histogram *n* a graphic representation of data consisting of a series of adjacent rectangles, the height and width of each rectangle being varied to represent each of 2 variables

histology *n* (anatomy that deals with) the organization and micro-

scopic structure of animal and plant tissues

histone *n* any of various proteins found associated with DNA in chromosomes

hives *n pl but sing or pl in constr* urticaria

hock *n* the tarsal joint of the hind limb of a horse or related quadruped that corresponds to the ankle in human beings

Hodgkin's disease *n* a malignant disease characterized by progressive anaemia with enlargement of the lymph glands, spleen, and liver

Holarctic *adj* of or being the biogeographical area that includes the northern parts of the Old World and New World

holdfast *n* a part by which an alga or other organism clings to a (flat) surface

hologamous *adj* having gametes which are very similar in size and structure to vegetative cells

holometabolous *adj, of an insect* having undergone complete metamorphosis

holophytic *adj* obtaining food in the manner of a green plant by synthesizing complex foods from simple inorganic substances

holothurian *n* any of a class of echinoderms having an elongated muscular body (eg the sea cucumber)

holozoic *adj* obtaining food in the manner of most animals by ingest-

hom

ing, digesting, and assimilating complex organic matter

hominid *n* any of a family of biped primate mammals comprising recent man and his immediate ancestors

hominoid *adj* resembling or related to man

Homo *n* any of a genus of primate mammals including recent man and various extinct ancestors

homoeostasis *n* the physiological maintenance of relatively constant conditions (eg constant internal temperature) within the body in the face of changing external conditions

homogen·ize, -ise *vt* to reduce the particles of so that they are uniformly small and evenly distributed; *esp* to break up the fat globules of (milk) into very fine particles ~ *vi* to become homogenized

homogenous *adj* of or exhibiting homogeny

homogeny *n* correspondence between parts or organs due to descent from the same ancestral type

homograft *n* a graft of tissue taken from a donor of the same species as the recipient

homoiotherm *n* a warm-blooded organism

homologous *adj* **1a** having the same relative position, value, or structure **b(1)** exhibiting biological homology **(2)** *of chromosomes* joining together with each other in pairs at meiotic cell division and having the same or corresponding genes **2** derived from an organism of the same species <*a ~ tissue graft*>

homologue, *NAm also* **homolog** *n* a chemical compound, chromosome, etc that exhibits homology

homology *n* correspondence in structure but not necessarily in function **a** between different parts of the same individual **b** between parts of different organisms due to evolutionary differentiation from a common ancestor

homomorphy *n* similarity of form (with different fundamental structure or origin)

homopterous *adj* of a large suborder of true bugs that have sucking mouthparts and include the aphids and cicadas

Homo sapiens *n* the binomial for mankind

homosexual *adj or n* (of, for, or being) sby having a sexual preference for members of his/her own sex — compare HETEROSEXUAL

homozygote *n* an animal, plant, or cell having identical alleles of a particular gene and so breeding true for that gene — compare HETEROZYGOTE

honeycomb *n* **1** (sthg resembling in shape or structure) a mass of 6-sided wax cells built by honeybees in their nest to contain their brood and stores of honey **2** (tripe from) the second stomach of a cow or other ruminant mammal

honeydew *n* a sweet deposit secreted on the leaves of plants usu by aphids

honey sac *n* a distension of the oesophagus of a bee in which honey is produced

hookworm *n* (infestation with or disease caused by) any of several parasitic nematode worms that have strong mouth hooks for attaching to the host's intestinal lining

horizon *n* any of the reasonably distinct soil or subsoil layers in a vertical section of land

hormone *n* (a synthetic substance with the action of) a product of living cells that usu circulates in body liquids (eg the blood or sap) and produces a specific effect on the activity of cells remote from its point of origin

horn *n* **1** any of the usu paired bony projecting parts on the head of cattle, giraffes, deer, and similar hoofed mammals and some extinct mammals and reptiles **2** a permanent solid pointed part consisting of keratin that is attached to the nasal bone of a rhinoceros **3** a natural projection from an animal (eg a snail or owl) resembling or suggestive of a horn **4** the tough fibrous material consisting chiefly of keratin that covers or forms the

horns and hooves of cattle and related animals, or other hard parts (eg claws or nails)

horny *adj* (made) of horn

horsetail *n* any of a genus of flowerless plants related to the ferns ⟶ EVOLUTION, PLANT

horticulture *n* the science and art of growing fruits, vegetables, and flowers

host *n* **1** a living animal or plant on or in which a parasite or smaller organism lives **2** an individual into which a tissue or part is transplanted from another

hot flush *n* a sudden brief flushing and sensation of heat, usu associated with an imbalance of endocrine hormones occurring esp at the menopause

hull¹ *n* **1** the outer covering of a fruit or seed **2** the calyx that surrounds some fruits (eg the strawberry)

hull² *vt* to remove the hulls of

humanoid *adj* having human form or characteristics

humerus *n*, *pl* **humeri** the long bone of the upper arm or forelimb extending from the shoulder to the elbow ⟶ ANATOMY

humify *vb* to convert into or form humus

humoral *adj* of or relating to a bodily fluid or secretion (eg an endocrine hormone)

humour, *NAm chiefly* **humor** *n* any of the 4 fluids of the body (blood, phlegm, and yellow and black bile) formerly held to determine, by their relative proportions, a person's health and temperament

humus *n* a brown or black organic soil material resulting from partial decomposition of plant or animal matter

husbandry *n* farming, esp of domestic animals

husk[1] *n* a dry or membranous outer covering (eg a shell or pod) of a seed or fruit

husk[2] *vt* to strip the husk from

hyaline *adj, of biological materials or structures* (nearly) transparent

hyaline cartilage *n* translucent bluish white cartilage that is present in joints and respiratory passages and forms most of the foetal skeleton

hyaloid *adj, of biological materials or structures* glassy, transparent

hyaloplasm *n* the clear, fluid, apparently homogeneous basic substance of cytoplasm

hybrid *n* 1 an offspring of 2 animals or plants of different races, breeds, varieties, etc 2 sthg heterogeneous in origin or composition

hybrid vigour *n* heterosis

hydatid *n* (a fluid-filled sac produced by and containing) a tape worm larva

hydra *n* any of numerous small tubular freshwater polyps having a mouth surrounded by tentacles

hydrocele *n* an accumulation of watery liquid in a body cavity (eg the scrotum)

hydrocephalus *also* **hydrocephaly** *n* an abnormal increase in the amount of cerebrospinal fluid within the brain cavity accompanied by enlargement of the skull and brain atrophy

hydrochloric acid *n* a solution of hydrogen chloride in water that is a strong corrosive acid and is naturally present in the gastric juice

hydrocortisone *n* a steroid hormone that is produced by the cortex of the adrenal gland and used esp in the treatment of rheumatoid arthritis

hydrolase *n* an enzyme concerned with hydrolysis

hydrolysis *n* chemical breakdown involving splitting of a bond and addition of the elements of water

hydrophobia *n* 1 abnormal dread of water 2 rabies

hydrophyte *n* a plant that grows in water or waterlogged soil

hydroponics *n pl but sing in constr* the growing of plants in (a mechanically supporting medium containing) nutrient solutions rather than soil

hydrotherapy *n* the use of water in the treatment of disease; *esp* treatment using exercise in heated water

hydrotropism *n* a tropism (eg in plant roots) in which water (vapour) is the orienting factor

hydrozoan *n* any of a class of

coelenterates that includes simple and compound polyps and jelly-fishes

hygrophilous *adj* living or growing in moist places

hymen *n* a fold of mucous membrane partly closing the opening of the vagina in virgins

hymenium *n, pl* **hymenia, hymeniums** a spore-bearing layer in fungi

hymenopteran, hymenopteron *n* any of an order of highly specialized usu stinging insects (eg bees, wasps, or ants) that often associate in large colonies and have usu 4 membranous wings

hyoid bone *n* a complex of joined bones situated at the base of the tongue and supporting the tongue and its muscles

hyperaemia *n* excess of blood in a body part

hyperglycaemia *n* excess of sugar in the blood (eg in diabetes mellitus)

hyperkinetic *adj* of or marked by abnormally increased, usu uncontrollable, muscular movement

hypermetropia *n* a condition in which visual images come to a focus behind the retina of the eye and vision is better for distant than for near objects; longsightedness — compare MYOPIA

hypersensitive *adj* abnormally susceptible (eg to a drug or antigen)

hypertension *n* (the systemic condition accompanying) abnormally high (arterial) blood pressure

hyperthermia *n* very high body temperature

hyperthyroidism *n* (the condition of increased metabolic and heart rate, enlargement of the thyroid gland, nervousness, etc resulting from) excessive activity of the thyroid gland — compare HYPOTHYROID-ISM

hypertonic *adj* **1** having excessive muscular tone or tension **2** having a higher concentration than a surrounding medium or a liquid under comparison — compare HYPO-TONIC, ISOTONIC

hypertrophy *n* excessive increase in bulk of an organ or part

hyperventilation *n* excessive breathing leading to abnormal loss of carbon dioxide from the blood

hypha *n, pl* **hyphae** any of the threads that make up the mycelium of a fungus

hypnogenesis *n* the induction of a hypnotic state

hypnosis *n, pl* **hypnoses** any of various conditions that (superficially) resemble sleep; *specif* one induced by a person to whose suggestions the subject is then markedly susceptible

hypnotherapy *n* the (psychotherapeutic) treatment of mental or physical disease, compulsive behaviour, etc using hypnosis

hypocotyl *n* the part of a plant embryo or seedling below the cotyledon

hyp

hypodermic[1] *adj* **1** of the parts beneath the skin **2** adapted for use in or administered by injection beneath the skin

hypodermic[2] *n* **1** a hypodermic injection **2** HYPODERMIC SYRINGE

hypodermic syringe *n* a small syringe used with a hollow needle for injection or withdrawal of material beneath the skin

hypogeal, hypogeous, hypogean *adj* growing, remaining, or occurring below the surface of the ground <~ *cotyledons*> — compare EPIGEAL

hypoglycaemia *n* abnormally low amount of sugar in the blood

hypogynous *adj* (having floral organs) attached to the receptacle or axis below the ovary and free from it — compare EPIGYNOUS, PERIGYNOUS

hypolimnion *n, pl* **hypolimnia** the (oxygen-deficient nutrient-rich) water below the thermocline of a lake — compare EPILIMNION

hypophysis *n, pl* **hypophyses** PITUITARY GLAND

hypoplasia *n* arrested development in which an organ or part remains below the normal size or in an immature state

hyposensit·ize, -ise *vt* to reduce the sensitivity of, esp to sthg that causes an allergic reaction; desensitize

hypotension *n* abnormally low blood pressure

hypothalamus *n* a part of the brain that lies beneath the thalamus and includes centres that regulate body temperature, appetite, and other autonomic functions

hypothermia *n* abnormally low body temperature

hypothyroidism *n* (the condition of lowered metabolic rate, lethargy, etc resulting from) deficient activity of the thyroid gland — compare HYPERTHYROIDISM

hypotonic *adj* **1** having deficient muscular tone or tension **2** having a lower concentration than a surrounding medium or a liquid under comparison — compare HYPERTONIC, ISOTONIC

hypoxia *n* a deficiency of oxygen reaching the tissues of the body

hysterectomy *n* surgical removal of the uterus

ichthyology *n* a branch of zoology that deals with fishes

ichthyophagous *adj* eating or living on fish

id *n* the one of the 3 divisions of the mind in psychoanalytic theory that is completely unconscious and is the source of psychic energy derived from instinctual needs and drives — compare EGO, SUPEREGO

identical *adj, of twins, triplets, etc* derived from a single egg

identification *n* **1** the putting of oneself mentally in the position of another **2** the psychological (unconscious) attribution of the characteristics of another to oneself in order to attain gratification, emotional support, etc

idiocy *n* extreme mental deficiency

idiopathic *adj, of a disease* arising spontaneously or from an unknown cause

ileum *n, pl* **ilea** the last division of the small intestine extending between the jejunum and the large intestine

ileus *n* obstruction of the bowel

iliac *also* **ilial** *adj* of or located near the ilium

ilium *n, pl* **ilia** the upper and largest of the 3 principal bones composing either half of the pelvis

imago *n, pl* **imagoes, imagines** an insect in its final mature (winged) state ⟿ LIFE CYCLE

imbalance *n* lack of functional balance in a physiological system ⟨hormonal ~⟩

imbricate[1] *adj* (having scales, sepals, etc) lying lapped over each other in regular order

imbricate[2] *vb* to overlap, esp in regular order

immature *adj* lacking complete growth, differentiation, or development

immigrant *n* a plant or animal that becomes established in an area where it was previously unknown

immobil·ize, -ise *vt* to reduce or eliminate motion of (sby or a body part) by mechanical means or by strict bed rest

immune *adj* **1** having a high degree of resistance to a disease ⟨~ *to diphtheria*⟩ **2** having or producing antibodies to a corresponding antigen ⟨*an* ~ *serum*⟩ **3** concerned with or involving immunity ⟨*an* ~ *response*⟩

immunity *n* being immune; *specif* the ability to resist the effects or development of a disease-causing parasite, esp a microorganism

immunoassay *n* the identification and measurement of the concentration of a substance (eg a protein) through its capacity to act as an antigen in the presence of specific antibodies that react with it

immunoglobulin *n* a protein (eg an antibody) that is made up of light and heavy amino acid chains and usu binds specifically to a particular antigen

immunology *n* biology that deals with the phenomena and causes of immunity

immunosuppression *n* suppression (eg by drugs) of natural immune responses

immunotherapy *n* treatment of or preventive measures against disease by administering (preparations of) antigens

imperfect *adj, of plants* not having the stamens and carpels in the same flower

impermeable *adj* not permitting passage, esp of a fluid

implant *vt* **1** to set permanently in

the consciousness or habit patterns **2** to insert in the tissue of a living organism

implantation *n* attachment of a mammalian embryo to the wall of the uterus ocurring prior to the formation of the placenta

impotent *adj* unable to copulate through an inability to maintain an erection of the penis

impregnate *vt* **1** to introduce sperm cells into **2** to make pregnant; fertilize

imprinting *n* a behaviour pattern rapidly established early in the life of an animal that involves attachment to an object or other animal, esp the animal's mother, seen just after birth

impulse *n* **1** a wave of excitation transmitted through a nerve that results in physiological (eg muscular) activity or inhibition **2** a sudden spontaneous inclination or incitement to some usu unpremeditated action **3** a propensity or natural tendency, usu other than rational

inarticulate *adj* not jointed or hinged

inborn *adj* hereditary, inherited

inbred *adj* subjected to or produced by inbreeding

inbreeding *n* the interbreeding of closely related individuals, esp to preserve and fix desirable characters

incisor *n* a tooth adapted for cut-

ting; *esp* one of the cutting teeth of mammals in front of the canines

incompatible *adj* unsuitable for use together because of undesirable chemical or physiological effects <~ *drugs*>

incomplete *adj* lacking a part

incontinent *adj* **1** lacking self-restraint (eg in sexual appetite) **2** suffering from lack of control of urination or defaecation

increase *vi* **1** to become progressively greater (eg in size, amount, quality, number, or intensity) **2** to multiply by the production of young ~ *vt* to make greater

incubate *vt* to sit on so as to hatch (eggs) by the warmth of the body; *also* to maintain (eg an embryo or a chemically active system) under conditions favourable for hatching, development, or reaction ~ *vi* **1** to sit on eggs **2** to undergo incubation

incubation *n* **1** incubating **2** the period between infection by a disease-causing agent and the manifestation of the disease

incubator *n* **1** an apparatus in which eggs are hatched artificially **2** an apparatus that maintains controlled conditions, esp for the housing of premature or sick babies or the cultivation of microorganisms

incus *n, pl* **incudes** the middle bone of a chain of 3 small bones in the

ear of a mammal; the anvil ➤
SENSE ORGAN

indehiscent *adj* remaining closed at maturity <~ *fruits*>

indicator *n* **1** a substance (eg litmus) used to show visually (eg by change of colour) the condition or contents of a solution **2** a tracer

indigenous *adj* originating, growing, or living naturally in a particular region or environment <~ *to Australia*>

indoleacetic acid *n* a plant hormone that promotes growth and rooting of plants

induce *vt* **1** to cause to appear or to happen; ; *specif* to cause (labour) to begin by the use of drugs **2** to cause the formation of

indusium *n, pl* **indusia** a covering outgrowth or membrane (eg of a cluster of fern spores)

industrial melanism *n* genetically determined darkening, esp in insects that occur in areas blackened by industrial pollutants

inert *adj* deficient in active (chemical or biological) properties

infantilism *n* **1** retention of childish physical, mental, or emotional qualities in adult life **2** an act or expression that indicates lack of maturity — used technically

infarct *n* an area of death in a tissue or organ resulting from obstruction of the local blood circulation

infect *vt* **1** to contaminate (eg air or food) with a disease-causing agent

2 to pass on a disease or a disease-causing agent to **3** to invade (an individual or organ), usu by penetration — used with reference to a pathogenic organism

infection *n* **1** infecting **2** (an agent that causes) a contagious or infectious disease

infectious *adj* **1** **infectious, infective** capable of causing infection **2** communicable by infection — compare CONTAGIOUS

infectious mononucleosis *n* an acute infectious disease characterized by fever and swelling of lymph glands

inferior *adj* **1** *of an animal or plant part* situated below or at the base of another (corresponding) part **2** *of a calyx* lying below the ovary **3** *of an ovary* lying below the petals or sepals

inferiority complex *n* a sense of personal inferiority often resulting either in timidity or, through over-compensation, in exaggerated aggressiveness

infertile *adj* not fertile or productive <~ *eggs*> <~ *fields*>

infest *vt* to live in or on as a parasite

inflammation *n* a response to cellular injury marked by local redness, heat, and pain

inflorescence *n* **1a** (the arrangement of flowers on) a floral axis **b** a flower cluster; *also* a solitary flower ➤ FLOWER **2** the budding and unfolding of blossoms; flowering

influenza n **1** a highly infectious virus disease characterized by sudden onset, fever, severe aches and pains, and inflammation of the respiratory mucous membranes **2** any of numerous feverish usu virus diseases of domestic animals marked by respiratory symptoms

infundibular, infundibulate adj **1** funnel-shaped **2** of or having an infundibulum

infundibulum n, pl **infundibula** the funnel-shaped mass of grey matter that connects the pituitary gland to the brain

ingest vt to take in (as if) for digestion; absorb

ingrowth n **1** a growing inwards **2** sthg that grows in or into a space

inguinal adj of or situated in the groin region

inherit vt to receive by genetic transmission <~ *a strong constitution*>

inheritable adj capable of being inherited

inheritance n **1a** the transmission of genetic information from parent to offspring **b** the acquisition of a possession, condition, or trait from past generations **2** sthg that is or may be inherited

inhibit vt to discourage from free or spontaneous activity, esp by psychological or social controls ~ vi to cause inhibition

inhibition n **1** a psychological restraint on another psychological or physical activity (eg of a bodily organ or enzyme)

inhibitor, inhibiter n sthg that slows down or interferes with a chemical action

inject vt to force a fluid into

injection n **1** injecting **2** sthg (eg a medication) that is injected

ink n the black secretion of a squid or similar cephalopod mollusc that hides it from a predator or prey

inkblot test n RORSCHACH TEST

innate adj **1** existing in or belonging to an individual from birth **2** ENDOGENOUS 2

inner ear n the innermost part of the ear from which sound waves are transmitted to the brain as nerve impulses

innervate vt to supply with nerves

innominate bone n the large bone composed of the ilium, ischium, and pubis that forms half of the pelvis in mammals; the hipbone

inoculate vt **1** to introduce a microorganism into <~ *mice with anthrax*> **2** to introduce (eg a microorganism) into a culture, animal, etc for growth **3** to vaccinate

inoculum n, pl **inocula** material used for inoculation

inoperable adj not suitable for surgery

input n an amount coming or put in

insanitary adj unclean enough to

endanger health; filthy, contaminated

insect n **1** any of a class of arthropods with a well-defined head, thorax, and abdomen, only 3 pairs of legs, and typically 1 or 2 pairs of wings ⊙ ☞ EVOLUTION **2** any of various small invertebrate animals (eg woodlice and spiders) — not used technically

insecticide n sthg that destroys insects

insectivore n **1** any of an order of mammals including moles, shrews, and hedgehogs that are mostly small, nocturnal, and eat insects ☞ MAMMAL **2** an insect-eating plant or animal

inseminate vt to introduce semen into the genital tract of (a female)

insensible adj **1** having lost consciousness **2** lacking or deprived of sensory perception <~ to pain>

insert vi, of a muscle to be in attachment to a specified part <muscles ~ on bone>

insertion n the mode or place of attachment of an organ or part

insomnia n prolonged (abnormal) inability to obtain adequate sleep

inspiration n the drawing of air into the lungs

inspirator n an injector, respirator, etc by which gas, vapour, etc is drawn in

inspire vt to inhale vi to breathe in

instability n lack of (emotional or mental) stability

instar n (an insect or similar arthropod in) a (particular) stage between successive moults

instinct n (a largely inheritable tendency of an organism to make a complex and specific) response to environmental stimuli without involving reason

insular adj of an island of cells or tissue

insulin n a protein pancreatic hormone secreted by the islets of Langerhans that is essential esp for the metabolism of carbohydrates and is used in the treatment of diabetes mellitus

integument n a skin, membrane, husk, or other covering or enclosure, esp of (part of) a living organism

intelligence quotient n a number expressing the ratio of sby's intelligence as determined by a test to the average for his/her age

intensive adj constituting or relating to a method designed to increase productivity by the expenditure of more capital and labour rather than by increase in the land or raw mc-erials used <~ farming>

interbreed vb interbred vi **1** to cross-breed **2** to breed within a closed population ~ vt to cause to interbreed

intercellular adj occurring between cells <~ spaces>

intercostal adj (of a part) situated between the ribs

113

intercourse n 1 connection or exchange between persons, or groups, 2 physical sexual contact between individuals that involves the genitals of at least 1 person <*oral ~*>; *esp* SEXUAL INTERCOURSE

intercrop vb **-pp-** vt to grow a crop in between rows, plots, etc of (another crop) ~ vi to grow 2 or more crops simultaneously on the same plot

intercross n (a product of) crossbreeding

interferon n a protein that inhibits the development of viruses and is produced by cells in response to infection by a virus

intermission n an intervening period of time (eg between attacks of a disease)

intermolecular adj existing or acting between molecules

internal adj 1 applied through the stomach by swallowing <*an ~ medicine*> 2 (present or arising) within (a part of) the body or an organism <*an ~ organ*><*an ~ stimulus*>

internode n an interval or part between 2 nodes (eg of a plant stem)

interoceptive adj of or being stimuli arising within the body, esp in the viscera

interphase n the interval between the end of one mitotic or meiotic division and the beginning of another

intersex n (the condition of being) an intersexual individual

intersexual adj intermediate in sexual characters between a typical male and a typical female

interspecific adj existing or arising between different species

intervertebral disc n any of the tough elastic discs between the bodies of adjoining vertebrae

intestinal adj of, being, affecting, or occurring in the intestine

intestine n the tubular part of the alimentary canal that extends from the stomach to the anus

intoxicate vt 1 ²POISON 1 2 to excite or stupefy by alcohol or a drug, esp to the point where physical and mental control is markedly diminished

intracellular adj situated, occurring, or functioning within a living cell <*~ enzymes*>

intractable adj not easily relieved or cured <*~ pain*>

intramuscular adj in or going into a muscle

intraspecific adj occurring within a species; involving members of 1 species

intrauterine adj situated, used, or occurring in the uterus

intravascular adj situated or occurring in a (blood) vessel

intravenous adj situated or occurring in, or entering by way of a

Parts of an insect: a bee

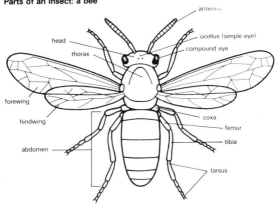

Mouth parts of a butterfly

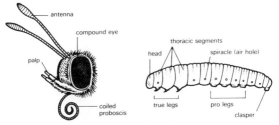

Caterpillar

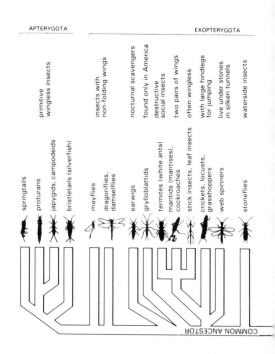

APTERYGOTA

EXOPTERYGOTA

primitive
wingless insects

insects with
non-folding wings

nocturnal scavengers

found only in America

destructive
social insects

two pairs of wings

often wingless

with large hindlegs
for jumping

live under stones
in silken tunnels

waterside insects

springtails

proturans

japygids, campodeids

bristletails (silverfish)

mayflies

dragonflies,
damselflies

earwigs

grylloblattids

termites (white ants)

mantids (mantises),
cockroaches

stick insects, leaf insects

crickets, locusts,
grasshoppers

web spinners

stoneflies

COMMON ANCESTOR

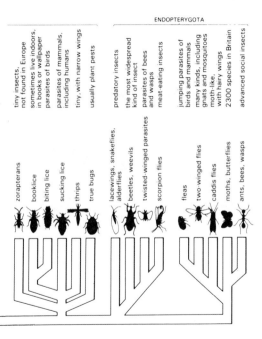

ENDOPTERYGOTA

zorapterans	tiny insects, not found in Europe
booklice	sometimes live indoors, in books or wallpaper
biting lice	parasites of birds
sucking lice	parasites of mammals, including humans
thrips	tiny, with narrow wings
true bugs	usually plant pests
lacewings, snakeflies, alderflies	predatory insects
beetles, weevils	the most widespread kind of insect
twisted-winged parasites	parasites of bees and wasps
scorpion flies	meat-eating insects
fleas	jumping parasites of birds and mammals
two-winged flies	many kinds, including gnats and mosquitoes
caddis flies	moth-like, with hairy wings
moths, butterflies	2300 species in Britain
ants, bees, wasps	advanced social insects

vein; *also* used in intravenous procedures

intrinsic *adj* originating or situated within the body

intrinsic factor *n* a substance produced by the lining of the intestines that is required for the absorption of vitamin B_{12} — compare EXTRINSIC FACTOR

introduction *n* sthg introduced; *specif* a plant or animal new to an area

introvert[1] *vt* to turn inwards or in on itself or oneself: eg **a** to bend inward; *also* to draw in (a tubular part) usu by invagination **b** to concentrate or direct (the mind, thoughts, or emotions) on oneself

introvert[2] *n* **1** sthg (eg the eyestalk of a snail) that is or can be drawn in **2** one whose attention and interests are directed towards his/her own mental life — compare EXTROVERT

intumescence *n* **1** an instance of enlarging or swelling **2** the state of being swollen **3** sthg swollen or enlarged

in utero *adv* in the uterus

invaginate *vt* to enclose. sheathe **2** to fold in so that an outer becomes an inner surface ~ *vi* to undergo invagination

invagination *n* **1** invaginating **2** an invaginated part

inversion *n* **1** a reversal of order, position, form, or relationship (eg of sections of a chromosome during chiasma formation)' **2** being turned inwards, inside out, or upside down

invertase *n* an enzyme capable of converting sucrose into invert sugar

invertebrate *n* any of that extensive group of animals that lacks a spinal column or notochord (eg protozoans, sponges, jellyfish, echinoderms, worms, molluscs, and arthropods)

invert sugar *n* a mixture of glucose and fructose found in fruits or produced artificially from sucrose

in vitro *adv or adj* outside the living body and in an artificial environment

in vivo *adv or adj* in the living body of a plant or animal

involucre *n* **1** or more whorls of bracts situated below and close to a flower (cluster) or fruit

involuntary *adj* **1** done contrary to or without choice **2** not subject to conscious control; reflex <~ *muscle*>

involute[1] *adj* **1** curled spirally **2** curled or curved inwards, esp at the edge <*an ~ leaf*>

involute[2] *vi* to return to a former condition <*after pregnancy the uterus ~s*>

iodopsin *n* a light-sensitive pigment in the retinal cones that is important in the perception of colour, esp in daylight vision

ion *n* an atom or group of atoms that

carries a positive or negative electric charge as a result of having lost or gained one or more electrons **2** a free electron or other charged subatomic particle

IQ *n* INTELLIGENCE QUOTIENT

iris *n*, *pl* **irises**, **irides** the opaque contractile diaphragm perforated by the pupil that forms the coloured portion of the eye ☞ SENSE ORGAN

irradiation *n* exposure to radiation (eg X rays or alpha rays)

irregular *adj* lacking symmetry or evenness <an ~ *flower*>

irrigate *vt* **1** to supply (eg land) with water by artificial means **2** to flush (eg an eye or wound) with a stream of liquid ~ *vi* to practise irrigation

irritable *adj* **1** capable of being irritated **2** (excessively) responsive to stimuli

irritant *n* sthg that irritates or excites

irritate *vt* to induce a response to a stimulus in or of ~ *vi* to cause or induce displeasure or anger

ischium *n*, *pl* **ischia** the rearmost and lowest of the 3 principal bones composing either half of the pelvis

islet of Langerhans *n* any of the groups of endocrine cells in the pancreas that secrete insulin

isoenzyme *n* an isozyme

isogamete *n* a gamete similar in form, size, or behaviour to another gamete with which it may fuse to form a zygote

isogamous *adj* having or involving fusion of isogametes

isogenic *adj* characterized by essentially identical genes <*identical twins are* ~>

isoleucine *n* an essential amino acid found in most proteins and essential to the diet of human beings

isomerous *adj* having an equal number of parts (eg ridges or markings); *esp, of a flower* having the members of each floral whorl equal in number

isomorphic *adj* having or involving structural similarity or identity

isopod *n* any of a large order of small crustaceans with eyes not borne on stalks and having 7 pairs of similar legs

isotonic *adj* having the same concentration as a surrounding medium or a liquid under comparison — compare HYPERTONIC, HYPOTONIC

isozyme *n* any of 2 or more chemically distinct but functionally similar enzymes

jaundice *n* an abnormal condition marked by yellowish pigmentation of the skin, tissues, and body fluids caused by the deposition of bile pigments

jaw *n* **1** either of 2 cartilaginous or bony structures that in most vertebrates form a framework above and below the mouth in which the teeth are set **2** any of various organs of invertebrates that per-

jaw

form the function of the vertebrate jaws

jawbone *n* the bone of an esp lower jaw

jejunum *n* the section of the small intestine between the duodenum and the ileum

jellyfish *n* a free-swimming marine coelenterate that has a nearly transparent saucer-shaped body and extendable tentacles covered with stinging cells ☞ EVOLUTION

jerk *n* **1** an involuntary spasmodic muscular movement due to reflex action **2** *pl* spasmodic movements due to nervous excitement

joint[1] *n* **1** a point of contact between 2 or more bones of an animal skeleton together with the parts that surround and support it **2** NODE **2 3** a part or space included between 2 articulations, knots, or nodes

joint[2] *adj* being a function of or involving 2 or more random variables <*a ~ probability density function*>

jugular *adj* **1a** of the throat or neck **b** of the jugular vein **2** *of a ventral fin of a fish* located on the throat

jugular vein, jugular *n* any of several veins of each side of the neck that return blood from the head

juice *n* **1** the extractable fluid contents of cells or tissues **2** *pl* the natural fluids of an animal body

Jungian *adj* (characteristic) of the psychoanalytical psychology of Carl Jung

jungle *n* an area overgrown with thickets or masses of (tropical) trees and other vegetation

Jurassic *adj or n* (of or being) the middle period of the Mesozoic era between the Cretaceous and the Triassic

juvenile[1] *adj* physiologically immature or undeveloped

juvenile[2] *n* a young individual resembling an adult of its kind except in size and reproductive activity

juvenile hormone *n* an insect hormone that controls maturation to the imago and plays a role in reproduction

karyokinesis *n*, *pl* **karyokineses** (the division of the nucleus that occurs in) mitotic cell division

karyotype *n* (the sum of the specific characteristics of) the chromosomes of a cell

keel *n* a projection (eg the breastbone of a bird) suggesting a keel

keratin *n* any of various fibrous proteins that form the chemical basis of nails, claws, and other horny tissue and hair

kernel *n* **1** the inner softer often edible part of a seed, fruit stone, or nut **2** a whole seed of a cereal

ketone body *n* a ketone or related chemical compound found in the blood and urine in abnormal

120

amounts in conditions of impaired metabolism (eg diabetes mellitus)

ketosis *n* an abnormal increase of ketone bodies in the body

key *n* **1** an arrangement of the important characteristics of a group of plants or animals used for identification **2** a samara

kidney *n* **1** either of a pair of organs situated in the body cavity near the spinal column that excrete waste products of metabolism in the form of urine **2** an excretory organ of an invertebrate

kinaesthesia, *NAm chiefly* **kinesthesia** *n* the sense of the position and movement of the joints of the body

kinesics *n* a systematic study of the relationship between bodily cues or movements (eg eye movement, blushes, or shrugs) and communication

kinesis *n, pl* **kineses** a movement that lacks directional orientation but depends on the intensity of the stimulation

kinetic *adj* of the motion (and associated forces and energy) of material bodies

kinetin *n* a plant growth substance that increases mitosis and callus formation

kingdom *n* any of the 3 primary divisions into which natural objects are commonly classified — compare ANIMAL KINGDOM, MINERAL KINGDOM, PLANT KINGDOM

kinin *n* any of various polypeptide hormones that are formed locally in the tissues and chiefly affect smooth muscle

kiss of life *n* artificial respiration in which the rescuer blows air into the victim's lungs by mouth-to-mouth contact

Klinefelter's syndrome *n* an abnormal condition in a man characterized by 2 X and 1 Y chromosomes, infertility, and smallness of the testicles

knee *n* **1** (the part of the leg that includes) a joint in the middle part of the human leg that is the articulation between the femur, tibia, and kneecap **2** a corresponding joint in an animal, bird, or insect

kneecap *n* a thick flat triangular movable bone that forms the front point of the knee and protects the front of the joint

knee jerk *n* an involuntary forward kick produced by a light blow on the tendon below the kneecap

knit *vb* **knit, knitted; -tt-** *vt* to cause to grow together <*time and rest will ~ a fractured bone*> ~ *vi* to grow together

knot *n* **1** a protuberant lump or swelling in tissue **2** (a rounded cross-section in timber of) the base of a woody branch enclosed in the stem from which it arises

knuckle *n* the rounded prominence formed by the ends of the 2 bones at a joint; *specif* any of the joints between the hand and the fingers or the finger joints closest to these

Krebs cycle *n* a sequence of reactions in the living organism which provide energy stored in phosphate bonds

krill *n* planktonic crustaceans and larvae that are the principal food of whalebone whales

kwashiorkor *n* severe malnutrition in infants and children that is caused by a diet high in carbohydrate and low in protein

label *vt* to distinguish (an element or atom) by using a radioactive or other distinguishable isotope to trace the progress of a chemical reaction or a biological process

labial *adj* of the lips or labia

labia majora *n* the outer fatty folds bounding the vulva

labia minora *n* the inner highly vascular largely connective-tissue folds bounding the vulva

labiate¹ *adj, of a plant corolla or calyx* having 2 unequal portions resembling lips

labiate² *n* a plant of the mint family

labile *adj* 1 readily open to change <*an emotionally ~ person*> 2 unstable <*a ~ mineral*>

labium *n, pl* **labia 1** any of the folds at the margin of the vulva — compare LABIA MAJORA, LABIA MINORA ☞ REPRODUCTION **2** the (lower) lip of a flower divided into 2 lip-like parts **3** a lower mouthpart

of an insect **4** a liplike part of various invertebrates

labrum *n, pl* **labra** an upper or front mouthpart of an arthropod

labyrinth *n* (the tortuous anatomical structure in) the ear or its bony or membranous part

lac *n* a resinous substance secreted by a scale insect

laceration *n* a torn and ragged wound

lachrymal, lacrimal *adj* **1** of or constituting the glands that produce tears **2** of or marked by tears

lachrymation, lacrimation *n* the (abnormal or excessive) secretion of tears

laciniate *adj* bordered with a fringe <*a ~ petal*>

lactase *n* an enzyme that hydrolyzes lactose and certain other sugars

lactate *vi* to secrete milk

lacteal¹ *adj* **1** consisting of, producing, or resembling milk **2a** conveying or containing a milky fluid **b** of the lacteals

lacteal² *n* any of the lymphatic vessels conveying chyle to the thoracic duct

lactic *adj* of milk

lactic acid *n* an organic acid, normally present in living tissue, and used esp in food and medicine and in industry

lactiferous *adj* **1** secreting or conveying milk **2** yielding a milky juice

lactose n a disaccharide sugar that is present in milk

lacuna n, pl **lacunae, lacunas** a small cavity in an anatomical structure

lacustrine adj of or occurring in lakes

laevorotatory adj turning towards the left or anticlockwise; esp rotating the plane of polarization of light to the left — compare DEXTROROTATORY

lagomorph n any of an order of gnawing mammals comprising the rabbits and hares ☞ MAMMAL

Lamarckism n a theory of organic evolution asserting that changes in the environment of plants and animals cause changes in their structure that are transmitted to their offspring

lamella n, pl **lamellae** also **lamellas** a thin flat scale, membrane, or part (eg a gill of a mushroom)

lamellibranch n, pl **lamellibranchs** any of a class of bivalve molluscs (eg clams, oysters, and mussels)

lamina n, pl **laminae, laminas** a thin plate, scale, layer, or flake

laminate¹ vi to separate into laminae

laminate² adj covered with or consisting of laminae

lamination n 1 a laminate structure 2 a lamina

lamprey n any of an order eel-like aquatic vertebrates that have a large sucking mouth with no jaws ☞ EVOLUTION

lancelet n any of a subphylum of small translucent marine animals related to vertebrates ☞ EVOLUTION

lanceolate adj shaped like a lance head; specif tapering to a point at the apex and sometimes at the base <~ leaves> ☞ PLANT

lanugo n soft downy hair; esp that covering the foetus of some mammals, including humans

large intestine n the rear division of the vertebrate intestine that is divided into caecum, colon, and rectum, and concerned esp with the resorption of water and formation of faeces

larva n, pl **larvae** 1 the immature, wingless, and often wormlike feeding form that hatches from the egg of many insects and is transformed into a pupa or chrysalis from which the adult emerges 2 the early form (eg a tadpole) of an animal (eg a frog) that undergoes metamorphosis before becoming an adult USE ☞ LIFE CYCLE

larvicide n an agent for killing larval pests

laryngeal n a nerve, artery, etc that supplies or is associated with the larynx

laryngectomy n surgical removal of (part of) the larynx

laryngitis n inflammation of the larynx

larynx n, pl **larynges, larynxes** the modified upper part of the trachea of air-breathing vertebrates that contains the vocal cords in human beings, most other mammals, and a few lower forms ☞ RESPIRATION

lassa fever *n* an acute severe often fatal virus disease of tropical countries

latent *adj* present but not manifest <*a ~ infection*>

latent period *n* 1 the incubation period of a disease 2 the interval between stimulation and response

lateral *adj* of the side; situated on, directed towards, or coming from the side

lateral line *n* a sense organ along the side of a fish sensitive to low vibrations

latex *n, pl* **latices, latexes** a milky usu white fluid that is produced by various flowering plants (eg of the spurge and poppy families)

laxative *n or adj* (a usu mild purgative) having a tendency to loosen or relax the bowels (to relieve constipation)

layer[1] *n* 1 a branch or shoot of a plant treated to induce rooting while still attached to the parent plant 2 a plant developed by layering

layer[2] *vt* to propagate (a plant) by means of layers ~ *vi , of a plant* to form roots where a stem comes in contact with the ground

leader *n* a main or end shoot of a plant

leaf[1] *n, pl* **leaves** 1a any of the usu green flat and typically broadbladed outgrowths from the stem of a plant that function primarily in food manufacture by photosynthesis **b** a modified leaf (eg a petal or sepal) 2a (the state of having) foliage <*in ~*> **b** the leaves of a plant (eg tobacco) as an article of commerce

leaf[2] *vi* to shoot out or produce leaves

leafage *n* FOLIAGE 1

leaf curl *n* a plant disease characterized by curling of the leaves

leaflet *n* 1 any of the divisions of a compound leaf 2 a small or young foliage leaf

leaf mould *n* a compost or soil layer composed chiefly of decayed vegetable matter

leafstalk *n* a petiole

learned *adj* acquired by learning <*~ versus innate behaviour patterns*>

learning *n* modification of a behavioural tendency by experience (eg exposure to conditioning)

leech *n* any of a class of flesh-eating or bloodsucking usu freshwater worms

left atrioventricular valve *n* BICUSPID VALVE

leg *n* a limb of an animal used esp for supporting the body and for walking: eg **a** (an artificial replacement for) either of the lower limbs of a human **b** any of the appendages on each segment of an arthropod (eg an insect or spider) used in walking and crawling

legionnaire's disease *n* a serious sometimes fatal infectious disease like pneumonia that is caused by a bacterium and often affects groups of closely associated people

legume *n* **1** the (edible) pod or seed of a leguminous plant **2** any of a large family of plants, shrubs, and trees having pods containing 1 or many seeds and including important food and forage plants (eg peas, beans, or clovers)

lens *n* the transparent lens-shaped or nearly spherical body in the eye that focuses light rays (eg on the retina) ⮕ SENSE ORGAN

lenticel *n* a pore in the stems of woody plants through which gases are exchanged between the atmosphere and the stem tissues

lenticular *adj* of a lens

lepidopteran *n* any of a large order of insects comprising the butterflies, moths, and skippers that are caterpillars in the larval stage and have 4 wings usu covered with minute overlapping and often brightly coloured scales when adult

lepidopterist *n* a specialist in the study of lepidopterans

lepidopteron *n, pl* **lepidoptera** *also* **lepidopterons** a lepidopteran

lesion *n* abnormal change in the structure of an organ or part due to injury or disease

lethargy *n* abnormal drowsiness

leucine *n* an amino acid found in most proteins and essential to the diet of human beings

leucocyte *n* WHITE BLOOD CELL

leucoma *n* a dense white opaque part in the cornea of the eye

leucotomy *n* a lobotomy

leukaemia *n* any of several usu fatal types of cancer that are characterized by an abnormal increase in the number of white blood cells in the body tissues, esp the blood, and occur in acute or chronic form

levator *n, pl* **levatores, levators** a muscle that serves to raise a part of the body — compare DEPRESSOR

ley *n* arable land used temporarily for hay or grazing

libido *n, pl* **libidos 1** emotional or mental energy derived in psychoanalytic theory from primitive biological urges **2** sexual drive

lice *pl of* LOUSE

lichen *n* **1** any of numerous complex plants made up of an alga and a fungus growing in symbiotic association on a solid surface (eg a rock or tree trunk) ⮕ EVOLUTION **2** any of several skin diseases characterized by raised spots

lid *n* the operculum in mosses

life *n, pl* **lives 1a** the quality that distinguishes a vital and functional being from a dead body **b** a state of matter (eg a cell or an organism) characterized by capacity for metabolism, growth, reaction to stimuli, and reproduction **2a** the sequence of physical and mental experiences that make up the existence of an individual **b** an aspect of the process of living <*the sex ~ of the frog*> **3** living beings (eg of a specified kind or environment) <*forest ~*>

life cycle *n* the series of stages in form and functional activity through which an organism, group, culture, etc passes during its lifetime

life expectancy *n* the expected length of sby's or sthg's life, based on statistical probability

life history *n* the changes through which an organism passes in its development from the primary stage to its natural death

life science *n* a science (eg biology, medicine, anthropology, or sociology) that deals with living organisms and life processes

lifetime *n* the length of time for which a person, living thing, subatomic particle, etc exists

ligament *n* a tough band of connective tissue forming the capsule round a joint or supporting an organ (eg the womb)

light¹ *n* **1a** (the sensation aroused by) sthg that makes vision possible by stimulating the sense of sight **b** an electromagnetic radiation in the wavelength range including infrared, visible, ultraviolet, and X rays; *specif* the part of this range that is visible to the human eye **2** a medium (eg a window) through which light is admitted

light² *adj* easily pulverized; crumbly <*~ soil*>

lignify *vb* to convert into or become wood or woody tissue

lignin *n* a substance that forms the (cementing material between the) woody cell walls of plants

ligulate *adj* **1** shaped like a strap <*the ~ corolla of a ray flower of a composite plant*> ⇒ PLANT **2** having ligules

ligule *n* an appendage on a foliage leaf and esp on the part of a blade of grass that forms a sheath round the stem

limb¹ *n* **1** any of the projecting paired appendages of an animal body used esp for movement and grasping but sometimes modified into sensory or sexual organs; *esp* a leg or arm of a human being **2** a large primary branch of a tree

limb² *n* the broad flat part of a petal or sepal furthest from its base

limbic *adj* of or being a group of structures in the brain, including the hypothalamus and hippocampus, that are concerned esp with emotion and motivation

liminal *adj* **1** of or at a sensory threshold **2** barely perceptible

limnology *n* the scientific study of physical, chemical, biological, etc

126

conditions in fresh waters (eg lakes)

limp *adj* lacking firmness

linctus *n* any of various syrupy usu medicated liquids used to relieve throat irritation and coughing

lingual *adj* 1 of or resembling the tongue 2 lying near or next to the tongue

linkage *n* the relationship between genes on the same chromosome that causes them to be inherited together

Linnaean, Linnean *adj* of or following the systematic methods of the Swedish botanist Linné who established the system of binomial nomenclature for all living things

linoleic acid *n* a liquid unsaturated fatty acid found in oils obtained from plants (eg linseed or peanut oil) and essential for mammalian nutrition

linolenic acid *n* a liquid unsaturated fatty acid found esp in drying oils (eg linseed oil) and essential for mammalian nutrition

lip *n* 1 either of the 2 fleshy folds that surround the mouth **2a** a fleshy edge or margin (eg of a wound) **b** a labium

lipase *n* an enzyme that accelerates the hydrolysis or synthesis of fats or the breakdown of lipoproteins

lipid *n* any of various substances that with proteins and carbohydrates form the principal structural components of living cells and that

include fats, waxes, and related and derived compounds

lipogenesis *n* the formation of fatty acids in the living body

lipophilic *adj* having an affinity for lipids (eg fats)

lipoprotein *n* a conjugated protein that is a complex of protein and lipid

lithophyte *n* a plant that grows on rock

litter 1 the uppermost slightly decayed layer of organic matter on the forest floor 2 a group of offspring of an animal, born at 1 birth

littoral¹ *adj* of or occurring on or near a (sea) shore

littoral² *n* a coastal region; *esp* the intertidal zone

liver *n* 1 a large vascular glandular organ of vertebrates that secretes bile and causes changes in the blood (eg by converting blood sugar into glycogen) 2 any of various large digestive glands of invertebrates

liver fluke *n* any of various trematode worms that invade and damage the liver of mammals, esp sheep

liverish *adj* suffering from liver disorder; bilious

liverwort *n* any of a class of plants related to and resembling the mosses but differing in reproduction and development ➔ EVOLUTION

lizard *n* any of a suborder of reptiles

◉ life cycle

Incomplete metamorphosis
grasshopper

nymph with wing buds

moults

nymph with wings absent

moults

eggs

adult with wings

Complete metamorphosis
swallowtail butterfly

larva

moults

pupa
in which the larva's body is rebuilt to form the imago

eggs

imago

Complete metamorphosis
mackerel

larva with yolk sac

larva loses yolk sac as mouth and gut form

egg

young fry

adult

Complete metamorphosis
newt

1 – 2 weeks tadpole with external gills

3 weeks tadpole with front legs

7 – 8 weeks tadpole with front and back legs

10 weeks adult with lungs

eggs laid on underwater plants

life cycle ☜

Development of the avian egg turkey

5 days
embryo visible with a network of blood vessels which absorb nutrients from the yolk

newly laid egg
with a large yolk that will nourish the growing chick

12 days
embryo attached to a shrinking yolk by a single connective stalk

15 days
various organs, particularly the eyes, can be seen clearly

23 days
chick is fully formed and absorbs the remaining yolk into its abdomen

29 days
newly hatched chick

Development of a marsupial mammal

After a six-week gestation, the kangaroo gives birth to a tiny undeveloped offspring, which crawls to its mother's pouch and attaches itself to a nipple. When sufficiently developed the young kangaroo makes excursions from the pouch, but will return to suckle until it is a year old.

birth

young kangaroo crawls to pouch

attachment at the nipple

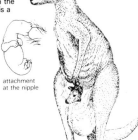

Development of a placental mammal

The young of placental mammals are nurtured and protected inside the mother during early development. Continuous nourishment and oxygen pass from the mother to the embryo via the placenta. As a result, the offspring are born at a more advanced stage of development than marsupial offspring.

loa

distinguished from the snakes by 2 pairs of well differentiated functional limbs (which may be lacking in burrowing forms), external ears, and eyes with movable lids

loam *n* a crumbly soil consisting of a mixture of clay, silt, and sand

lobe *n* a curved or rounded projection or division; *esp* such a projection or division of a bodily organ or part

lobotomy *n* a brain operation used, esp formerly, in the treatment of some mental disorders (eg violent psychoses) in which nerve fibres in the cerebral cortex are cut in order to change behaviour

lobule *n* (a subdivision of) a small lobe

local *adj* involving or affecting only a restricted part of a living organism

lockjaw *n* an early symptom of tetanus characterized by spasm of the jaw muscles and inability to open the jaws; *also* tetanus

locomotive *adj* of or functioning in locomotion

locomotor *adj* 1 locomotive 2 affecting or involving the locomotive organs

loculus *n, pl* loculi a small chamber or cavity, esp in a plant or animal body

locus *n, pl* loci *also* locuses the position on a chromosome of a particular gene or allele

loess *n* a usu yellowish brown loamy deposit found in Europe, Asia, and N America and believed to be chiefly deposited by the wind

loin *n* 1 the part of a human being or quadruped on each side of the spinal column between the hipbone and the lower ribs 2 *pl* **a** the upper and lower abdominal and the hip regions **b(1)** the pubic region (2) the genitals

loment *n* a dry 1-celled fruit that breaks transversely into numerous usu 1-seeded segments at maturity

longsighted *adj* hypermetropic

loop of Henle *n* a part of each nephron in a kidney that plays a part in water resorption

lorica *n, pl* **loricae** a hard protective case or shell

louse *n, pl* **lice 1** any of various small wingless usu flattened insects parasitic on warm-blooded animals **2** any of several small arthropods that are not parasitic — usu in combination <*book* ~ > <*wood* ~>

lower *adj* less advanced in the scale of evolutionary development <~ *organisms*>

luciferase *n* an enzyme that catalyses the oxidation of luciferin

luciferin *n* a protein in some organisms (eg fireflies and glowworms) that gives out practically heatless light when undergoing oxidation

lumbago *n* muscular pain of the lumbar region of the back

lumbar *adj* of or constituting the

loins or the vertebrae between the thoracic vertebrae and sacrum <*the ~ region*>

lumen *n, pl* **lumina, lumens** the cavity of a tubular organ <*the ~ of a blood vessel*>

luminesce *vi* to exhibit luminescence

luminescence *n* (an emission of) light that occurs at low temperatures and that is produced by physiological processes (eg in the firefly), by chemical action, by friction, or by electrical action

lung *n* **1** either of the usu paired compound saclike organs in the chest that constitute the basic respiratory organ of air-breathing vertebrates ☞ RESPIRATION **2** any of various respiratory organs of invertebrates

luteinizing hormone *n* a hormone from the front lobe of the pituitary gland that in the female stimulates the development esp of corpora lutea and in the male interstitial tissue of the testis

luteotrophic hormone *n* prolactin

luteotrophin, luteotropin *n* prolactin

lymph *n* a pale fluid resembling blood plasma that contains white blood cells but normally no red blood cells, that circulates in the lymphatic vessels, and bathes the cells of the body

lymphatic[1] *adj* **1** of, involving, or produced by lymph, lymphoid tissue, or lymphocytes **2** conveying lymph <*~ vessels*>

lymphatic[2] *n* a vessel that contains or conveys lymph

lymph gland *n* LYMPH NODE

lymph node *n* any of the rounded masses of lymphoid tissue that occur along the course of the lymphatic vessels and in which lymphocytes are formed

lymphocyte *n* a white blood cell that is present in large numbers in lymph and blood and defends the body by immunological responses to invading or foreign matter (eg by producing antibodies) — compare MONOCYTE

lymphoid *adj* **1** of or resembling lymph **2** of or constituting the tissue characteristic of the lymph nodes

lymphoma *n, pl* **lymphomas, lymphomata** a tumour of lymphoid tissue

lyrate *adj* shaped like a lyre ☞ PLANT

lysate *n* a product of lysis

lyse *vb* to (cause to) undergo lysis

lysin *n* a substance capable of causing lysis; *esp* an antibody capable of causing disintegration of red blood cells or microorganisms

lysine *n* a basic amino acid that is essential to nutrition in humans

lysis *n, pl* **lyses** **1** the gradual decline of a disease process (eg fever) **2** a process of disintegration or dissolution (eg of cells)

lys

lysosome *n* a vesicle surrounded by a membrane that occurs in cell cytoplasm and contains enzymes capable of breaking down unwanted material or causing autolysis

lysozyme *n* an enzyme present in egg white and in human tears and saliva that destroys the capsules of various bacteria

macrocephalous, macrocephalic *adj* having or being an exceptionally large head or cranium

macrogamete *n* the larger usu female gamete of a heterogamous organism

macromolecule *n* a large molecule (eg of a protein) built up from smaller chemical structures

macronutrient *n* a nutrient element of which relatively large quantities are essential to the growth and welfare of a plant — compare TRACE ELEMENT

macrophage *n* any of various large cells that are distributed throughout the body tissues, ingest foreign matter and debris, and may be attached to the fibres of a tissue or mobile

macrostructure *n* the structure of a metal, body part, the soil, etc revealed by visual examination with little or no magnification

macula lutea *n, pl* **maculae luteae** a small yellowish area lying slightly to the side of the centre of the retina that constitutes the region of best vision

maculation *n* the arrangement of spots and markings on an animal or plant

maggot *n* a soft-bodied legless grub that is the larva of a 2-winged fly (eg the housefly)

malabsorption *n* the deficient absorption of food substances, vitamins, etc (eg vitamin B_{12}) from the stomach and intestines

malacology *n* a branch of zoology dealing with molluscs

maladjusted *adj* poorly or inadequately adjusted, specif to one's social environment and conditions of life

malady *n* an animal disease or disorder

malaise *n* an indeterminate feeling of debility or lack of health, often accompanying the start of an illness

malaria *n* a disease caused by protozoan parasites in the red blood cells, transmitted by the bite of mosquitoes, and characterized by periodic attacks of chills and fever

male¹ *adj* **1** of or being the sex that produces relatively small sperms, spermatozoids, or spermatozoa by which the eggs of a female are fertilized **2** *of a plant or flower* having stamens but no ovaries

male² *n* a male person, animal, or plant

malformation *n* anomalous, abnor-

132

mal, or faulty formation or structure

malfunction *vi* to function imperfectly, badly, or not in the normal manner

malic acid *n* an acid found in the juices of certain fruits (eg apples) and other plant juices and is an intermediate in the Krebs Cycle

malignant *adj of a disease* very severe or deadly <~ *malaria*>; *specif, of a tumour* tending to infiltrate, spread, and cause death

malinger *vi* to pretend illness or incapacity so as to avoid duty or work

malleus *n, pl* **mallei** the outermost of the chain of 3 small bones that transmit sound to the inner ear of mammals; the hammer

malm *n* a soft crumbly limestone (soil)

malnutrition *n* faulty or inadequate nutrition

Malpighian body *n* the part of the nephron of mammals consisting of the glomerulus and its membrane

Malpighian tubule *n* any of a group of long vessels that open into the alimentary canal in insects and other arthropods and function esp as excretory organs

Malthusian *adj* of Malthus or his theory that population tends to increase faster than its means of subsistence and that widespread poverty inevitably results unless population growth is checked

maltose *n* a sugar formed esp from starch by amylase

mamillary, *NAm* **mammillary** *adj* of or resembling the breasts

mamma *n, pl* **mammae** a mammary gland with its accessory parts

mammal *n* any of a class of higher vertebrates comprising humans and all other animals that have mammary glands and nourish their young with milk ⊚ ⌖EVOLUTION

mammary *adj* of, lying near, or affecting the mammary glands

mammary gland *n* the breasts or other large compound modified skin glands in female mammals that secrete milk and are situated on the front of the body in pairs

man *n, pl* **men 1** a human being; *esp* an adult male as distinguished from a woman or child **2** the human race **3** a member of a family of biped primate mammals anatomically related to the great apes but distinguished esp by greater brain development and a capacity for articulate speech and abstract reasoning; *broadly* any ancestor of modern man

mandible *n* **1a** JAW 1 **b** a lower jaw together with its surrounding soft parts ⌖ ANATOMY **c** the upper or lower part of a bird's bill **2** any of various mouth parts in insects or other invertebrates for holding or biting food

mange *n* any of various contagious

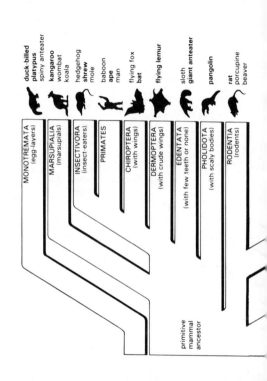

MONOTREMATA (egg-layers)
- duck-billed **platypus**
- spiny anteater

MARSUPIALIA (marsupials)
- **kangaroo**
- wombat
- koala

INSECTIVORA (insect-eaters)
- hedgehog
- **shrew**
- mole

PRIMATES
- baboon
- **ape**
- man

CHIROPTERA (with wings)
- flying fox
- **bat**

DERMOPTERA (with crude wings)
- **flying lemur**

EDENTATA (with few teeth or none)
- sloth
- **giant anteater**

PHOLIDOTA (with scaly bodies)
- **pangolin**

RODENTIA (rodents)
- **rat**
- porcupine
- beaver

primitive mammal ancestor

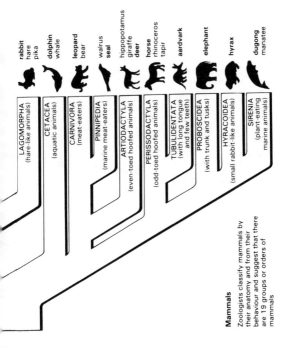

Mammals

Zoologists classify mammals by their anatomy and from their behaviour and suggest that there are 19 groups or orders of mammals

man

skin diseases affecting domestic animals or sometimes human beings, marked by inflammation and loss of hair and caused by a minute parasitic mite

mania *n* abnormal excitement and euphoria marked by mental and physical hyperactivity and disorganization of behaviour

manic-depressive *adj* of or affected by a mental disorder characterized by alternating mania and (extreme) depression

mantle *n* **1** a fold of a tunicate's, barnacle's, or mollusc's body wall (lining the shell) **2** the feathers covering the back, shoulders,and wings of a bird

manubrium *n, pl* **manubria** *also* **manubriums** the section of the sternum nearest the head of human beings and many other mammals

manure[1] *vt* to enrich (land) by the application of manure

manure[2] *n* material that fertilizes land; *esp* excreta from farmyard animals

maquis *n, pl* **maquis** (an area of) thick scrubby underbrush of Mediterranean shores

marasmus *n* progressive emaciation, esp in the young, due usu to faulty digestion and absorption of food

mariculture *n* the cultivation of marine organisms by exploiting their natural environment

marine *adj* of or (living) in the sea

Markov chain *n* a random sequence of states in which the probability of occurrence of a future state depends only on the present state and not on the path by which it was reached

marl *vt or n* (to fertilize with) a crumbly earthy deposit (eg of silt or clay) that contains calcium carbonate and is used esp as a fertilizer for lime-deficient soils

marlite *n* a marl resistant to the action of air

marram grass *n* any of several strong wiry grasses that grow on sandy shores and prevent erosion

marrow *n* **1** a soft tissue that fills the cavities and porous part of most bones and contains many blood vessels **2** the substance of the spinal cord

marrowbone *n* a bone rich in marrow

marsh *n* (an area of) soft wet land usu covered with sedges, rushes, etc

marsupial[1] *adj* **1** of or being a marsupial **2** of or forming a marsupium or pouch

marsupial[2] *n* any of an order of lower mammals including the kangaroos, wombats, and opossums that have a pouch on the abdomen of the female for carrying young and do not develop a placenta → LIFE CYCLE, MAMMAL

marsupium *n, pl* **marsupia** the abdominal pouch of a marsupial, formed by a fold of the skin and enclosing the mammary glands

mastectomy n excision or amputation of a breast

masticate vt to grind or crush (food) before swallowing, (as if) with the teeth; to chew

mastitis n inflammation of the breast or udder, usu caused by infection

mastoid adj or n (of, near, or being) a somewhat conical part of the temporal bone lying behind the ear

mate[1] vt 1 to join or fit together; couple 2a to join together as mates b to provide a mate for ~ vi to copulate

mate[2] n one of a pair; esp either member of a breeding pair of animals

matrix n, pl **matrices, matrixes** the substance between the cells of a tissue that holds them together

mature[1] adj having completed natural growth and development; ripe

mature[2] vt to bring to full development or completion ~ vi to become mature

maw n 1 an animal's stomach or crop 2 the throat, gullet, or jaws, esp of a voracious flesh-eating animal

maxilla n, pl **maxillae, maxillas 1a** JAW **b** (either of 2 bones of) the upper jaw of a human or other higher vertebrate 2 any of the (1 or 2 pairs of) mouthparts behind the mandibles in insects and other arthropods

meadow n (an area of moist low-lying usu level) grassland

mealybug n any of numerous scale insects with a white powdery covering that are pests, esp of fruit trees

mean[1] n a value that lies within a range of values and is computed according to a prescribed law; esp ARITHMETIC MEAN

mean[2] adj being the mean of a set of values <~ temperature>

measles n pl but sing or pl in constr 1 (German measles or another disease similar to) an infectious virus disease marked by a rash of distinct red circular spots 2 infestation with larval tapeworms, esp in pigs or pork

measly adj 1 infected with measles 2 containing larval tapeworms <~ pork>

mechanism n the physical or chemical processes involved in a natural phenomenon (eg an action, reaction, or biological evolution)

meconium n a dark greenish mass that accumulates in the bowels during foetal life and is discharged shortly after birth

media n, pl **mediae** the middle muscular part of the wall of a blood or lymph vessel

medial adj being, occurring in, or extending towards the middle; median

med

median[1] *n* **1** a median vein, nerve, etc **2** a value in a series above and below which there are an equal number of values

median[2] *adj* lying in the plane that divides an animal into right and left halves

medicine *n* **1** a substance or preparation used (as if) in treating disease **2** the science and art of the maintenance of health and the prevention and treatment of disease (using nonsurgical methods)

medium *n, pl* **mediums, media 1** a surrounding or enveloping substance **2** a nutrient for the artificial cultivation of bacteria and other (single-celled) organisms

medulla *n, pl* **medullas** *also* **medullae 1a** MARROW 1 **b** MEDULLA OBLONGATA **2a** the inner or deep part of an animal or plant structure <*the adrenal* ~> **b** the myelin sheath that surrounds some nerves

medulla oblongata *n, pl* **medulla oblongatas, medullae oblongatae** the (pyramid-shaped) part of the brain of vertebrates whose back part merges with the spinal cord

medullary *adj* **1** of or located in a medulla, esp the medulla oblongata **2** of or located in the pith of a plant

medullary ray *n* a wedge of tissue that is composed of parenchyma cells, joins the vascular bundles in the stems of many plants, and connects the pith with the cortex

medullated *adj* myelinated

medusa *n, pl* **medusae, medusas** a jellyfish

megagamete *n* a macrogamete

megagametophyte *n* the female gametophyte produced by a megaspore

megasporangium *n* a sporangium that produces only megaspores

megaspore *n* a spore that gives rise to a female gametophyte and is usu larger than a microspore

megasporophyll *n* a sporophyll that bears only megasporangia

meiosis *n, pl* **meioses** a specialized cellular process of nuclear division in gamete-producing cells by which 1 of each pair of homologous chromosomes passes to each resulting gametic cell which thus has half the number of chromosomes of the original cell — compare MITOSIS

melancholia *n* feelings of extreme depression and worthlessness occurring as an abnormal mental condition

melanin *n* a dark brown or black animal and plant pigment (eg of skin or hair)

melanism *n* an increased amount of (nearly) black pigmentation of skin, feathers, hair, etc

melanocyte-stimulating hormone *n* a hormone of the pituitary gland in vertebrates that produces darkening of the skin — compare MELATONIN

melanoma n, pl **melanomas** also **melanomata** a usu malignant tumour, esp of the skin, containing dark pigment

melanophore n a melanin-containing chromatophore, esp of fishes, amphibians, and reptiles

melanosis n the (abnormal) deposition of pigments, esp melanin, in the tissues of the body

melatonin n a hormone of the pineal gland in vertebrates that produces lightening of the skin — compare MELANOCYTE-STIMULATING HORMONE

membrane n a thin pliable sheet or layer, esp in an animal or plant

memoz! n (the power or process of recalling or realizing) the store of things learned and retained from an organism's experience <*good visual* ~>

menarche n (the onset of the menstrual function marked by) the first menstrual period

Mendelian adj of or according with the genetic principle of Gregor Mendel that genes occur in pairs, each gamete receives 1 member of each pair, and that an organism thus has 1 gene of each pair randomly selected from each of its parents

Ménière's disease n recurrent attacks of dizziness, ringing in the ears, and deafness occurring as a disorder of the inner ear

meninges pl of MENINX

meningitis n bacterial, fungal, or viral inflammation of the meninges

meninx n, pl **meninges** any of the 3 membranes (the dura mater, pia mater, and arachnoid) that envelop the brain and spinal cord — usu pl

menopause n (the time of) the natural cessation of menstruation occurring usu between the ages of 45 and 50

menorrhagia n abnormally profuse menstrual flow

menstruation n the discharging of blood, secretions, and tissue debris from the uterus that recurs in nonpregnant primate females of breeding age at approximately monthly intervals; also a single occurrence of this

mental adj 1 of the mind or its activity <~ *health*> <~ *processes*> 2 of, being, or (intended for the care of people) suffering from a psychiatric disorder <*a* ~ *patient*> <~ *illness*>

mental deficiency n failure in development of the mind resulting in a need for continuing parental or institutional care

mentality n mental power or capacity; intelligence

meristem n a plant tissue that is the major area of growth and is made up of small cells capable of dividing indefinitely

mesencephalon n the midbrain

mes

mesentery *n* any of several membranous double folds of the peritoneum of vertebrates, that envelop the intestines and connected organs and join them with the rear wall of the abdominal cavity

mesial *adj* (in or directed towards the) middle; *esp, of a plane* dividing an animal into right and left halves

mesoderm *n* (tissue derived from) the middle of the 3 primary germ layers of an embryo that is the source of bone, muscle, connective tissue, and the inner layer of the skin in the adult — compare ENDODERM, ECTODERM 2

mesoglea, mesogloea *n* a gelatinous substance between the ectoderm and endoderm of coelenterates

mesomorph *n* an individual with a body characterized by strong bone and muscular development

mesophyll *n* the parenchymatous tissue between the epidermal surface layers of a foliage leaf

mesophyte *n* a plant that grows under medium conditions of moisture

messenger RNA *n* an RNA that carries the code for the synthesis of a particular protein and acts as a template for its formation — compare TRANSFER RNA

metabolism *n* all the processes (by which a specified substance is dealt with) in the building up and destruction of living tissue; *specif* the chemical changes in living cells by which energy is provided and new material is assimilated

metabolite *n* **1** a product of metabolism **2** a substance essential to the metabolism of a particular organism or to a particular metabolic process

metacarpal *n* a bone of the metacarpus ☞ ANATOMY

metacarpus *n* the part of the hand or forefoot between the wrist and fingers that typically contains five elongated bones

metamorphose *vi* to undergo metamorphosis

metamorphosis *n, pl* **metamorphoses** a marked (abrupt) change in the form or structure of an insect, frog, etc occurring in the course of development ☞ LIFE CYCLE

metaphase *n* the stage of mitotic or meiotic cell division in which the chromosomes become arranged in the equatorial plane of the spindle

metaplasia *n* (abnormal) replacement of cells of one type by cells of another

metatarsal *n* a bone of the metatarsus ☞ ANATOMY

metatarsus *n* the part of the foot in human beings or of the hind foot in 4-legged animals between the ankle and toes

metaxylem *n* part of the primary xylem that differentiates after the protoxylem

metazoan *n* any of a kingdom or

140

subkingdom of animals that comprises all those with multicellular bodies differentiated into tissues

methionine *n* a sulphur-containing amino acid that is found in most proteins and is an essential constituent of human diet

microbe *n* a microorganism, germ

microbiology *n* the biology of bacteria and other microscopic forms of life

microcephalic *n or adj* (sby) having an abnormally small head and usu mental defects

microgamete *n* the smaller and usu male gamete of a heterogamous organism

micrograph *n* a drawing of the image of an object formed by a microscope

microhabitat *n* a small usu specialized and isolated habitat (eg a decaying tree stump)

micronutrient *n* a nutrient (eg a trace element) required in small quantities

microorganism *n* an organism of (smaller than) microscopic size

microphage *n* a small phagocyte

microphyll *n* a small leaf, esp of a club moss

micropyle *n* **1** a differentiated area of the surface of an egg through which the sperm enters **2** an opening in the surface of an ovule of a flowering plant through which the pollen tube penetrates, visible in the mature seed as a minute pore in the seed coat

microscope *n* an instrument consisting of (a combination of) lenses for making enlarged images of minute objects using light or other radiations

microscopic *also* **microscopical** *adj* **1** of or conducted with the microscope or microscopy **2** invisible or indistinguishable without the use of a microscope

microsome *n* a minute particle **a** seen in the cytoplasm of a cell viewed through a light microscope **b** seen in a fraction obtained by heavy centrifugation of broken cells viewed through an electron microscope

microsporangium *n* a sporangium that produces only microspores

microspore *n* a spore that gives rise to a male gametophyte and is usu smaller than a megaspore

microsporophyll *n* a sporophyll that develops only microsporangia

microstructure *n* the microscopic structure of a mineral, alloy, living cell, etc

microtome *n* an instrument for cutting very thin sections (eg of plant or animal tissues) for microscopic examination

micturate *vi* to (want to) urinate — *fml; sometimes used technically*

midbrain *n* (the parts of the adult brain that develop from) the mid-

mid

dle of the 3 primary divisions of the embryonic vertebrate brain

middle ear *n* a cavity through which sound waves are transmitted by a chain of tiny bones from the eardrum to the inner ear

midrib *n* the central vein of a leaf

migraine *n* recurrent severe headache usu associated with disturbances of vision, sensation, and movement often on only 1 side of the body

migrant *n* an animal that moves from one habitat to another

migrate *vi, of an animal* to pass usu periodically from one region or climate to another for feeding or breeding

mildew *n* (a fungus producing) a usu whitish growth on the surface of organic matter (eg paper or leather) or living plants

milk *n* **1** a (white or creamy) liquid secreted by the mammary glands of females for the nourishment of their young (and used as a food by humans) **2** a milklike liquid: eg **a** the latex of a plant **b** the juice of a coconut

milk fever *n* **1** a feverish disorder following childbirth **2** a disease of cows, sheep, goats, etc that have recently given birth, caused by a drain on the body's mineral reserves during the establishment of the milk flow

milk sugar *n* lactose

millet *n* (the seed of) any of various small-seeded annual cereal and forage grasses cultivated for their grain, used as food

millipede, millepede *n* any of a class of myriopods usu with a cylindrical segmented body and 2 pairs of legs on each segment ☞ EVOLUTION

mimesis *n* imitation, mimicry

mimetic *adj* relating to, characterized by, or exhibiting mimicry

mimic *vt* **-ck-** to resemble by biological mimicry

mimicry *n* resemblance of one organism to another or to natural objects in its environment that secures it an advantage (eg protection from predation)

mind *n* **1** the (capabilities of the) organized conscious and unconscious mental processes of an organism that result in reasoning, thinking, perceiving, etc **2** the normal condition of the mental faculties <*lost his* ~> **3** the intellect and rational faculties as contrasted with the emotions

mineral kingdom *n* that one of the 3 basic groups of natural objects that includes all inorganic objects

Miocene *adj or n* (of or being) an epoch of the Tertiary between the Pliocene and the Oligocene

miscarriage *n* the expulsion of a human foetus before it is viable, esp after the 12th week of gestation

miscarry *vi* to suffer miscarriage of a foetus

142

missing link *n* a supposed intermediate form between man and his anthropoid ancestors

mite *n* 1 any of an order of (extremely) small arachnids that often infest animals, plants, and stored foods

mitochondrion *n, pl* **mitochondria** any of several organelles in a cell that are rich in fats, proteins, and enzymes and produce energy through cellular respiration

mitosis *n, pl* **mitoses** the formation of 2 new nuclei from an original nucleus, each having the same number of chromosomes as the original nucleus, during cell division; *also* cell division in which this occurs — compare MEIOSIS

mitral valve *n* BICUSPID VALVE

mixed farming *n* the growing of food crops and the rearing of livestock on the same farm

mobil·ize, -ise *vb* to release (sthg stored in the body) for use in an organism

mode *n* the most frequently occurring value in a set of data

molar *n* one of the cheek teeth in mammals adapted for grinding ☞ ANATOMY

molecular biology *n* the study of the basic molecular organization and functioning of living matter

mollusc, *NAm chiefly* **mollusk** *n* any of a large phylum of invertebrate animals with soft bodies not divided into segments and usu enclosed in a shell, including the snails, shellfish, octopuses, and squids ☞ ANATOMY

molt *vb or n, NAm* (to) moult

monadelphous *adj, of stamens* united by the filaments into 1 group usu forming a tube around the carpels

monandrous *adj* having (flowers with) a single stamen

monandry *n* a monandrous condition of a plant or flower

monecious *adj, NAm* monoecious

mongolism *n* DOWN'S SYNDROME

monocarpic *adj, of a plant* bearing fruit only once and then dying

monochlamydeous *adj* having only one whorl of perianth segments

monocotyledon *n* any of a subclass of plants comprising all those with a single cotyledon and usu parallel-veined leaves (eg the grasses, orchids, and lilies)

monoculture *n* the cultivation of a single agricultural product to the exclusion of other uses of the land

monocyte *n* a large white blood cell that is present in small numbers in the blood and defends the body by engulfing and digesting invading or unwanted matter — compare LYMPHOCYTE

monoecious, *NAm also* **monecious** *adj* hermaphroditic; *esp* having female and male flowers on the same plant — compare DIOECIOUS

monoestrous *n* experiencing oestrus once each year; having a single annual breeding period

monogerm *adj* producing or being a fruit that gives rise to a single plant <*a ~ variety of sugar beet*>

monohybrid *n or adj* (an organism, cell, etc) having 2 different versions of 1 specified gene

monosaccharide *n* a sugar (eg glucose) not decomposable to simpler sugars

monotreme *n* any of an order of lower mammals comprising the platypus and echidna ☞ MAMMAL

monozygotic *adj*, *of twins, triplets, etc* identical

montane *adj* of, being, or growing or living in the area of cool slopes just below the tree line on mountains

moor *n* an expanse of open peaty infertile usu heath-covered upland

mor *n* a humus usu that forms a distinct layer above the underlying soil

morbidity *n* the relative incidence of (a) disease

morphogenesis *n* the formation and differentiation of tissues and organs (during embryonic development)

morphology *n* (the biology of) the form and structure of animals and plants

mortality *n* **1** being mortal **2** the number of deaths in a given time or place

morula *n* an embryo during the process of cleavage preceding the blastula stage, consisting of a solid globular mass of cells

mosaic *n* **1** (a part of) an organism composed of cells with different genetic make-up; chimera **2** a virus disease of plants (eg tobacco) characterized esp by diffuse yellow and green mottling of the foliage

mosquito *n, pl* **mosquitoes** *also* **mosquitos** any of a family of 2-winged flies with females that suck the blood of animals and often transmit diseases (eg malaria) to them

moss *n* (any of various plants resembling) any of a class of primitive plants with small leafy stems bearing sex organs at the tip; *also* many of these plants growing together and covering a surface ☞ EVOLUTION, PLANT

moss animal *n* a bryozoan

mother *adj* acting as or providing a parental stock — used without reference to sex

motile *adj* exhibiting or capable of movement

motor *adj* **1** of or being a nerve (fibre) that conducts an impulse causing the movement of a muscle **2** of or involving muscular movement

mould¹, *NAm chiefly* **mold** *n* crumbling soft (humus-rich) soil suited to plant growth

mould², *NAm chiefly* **mold** *n* (a

fungus producing) an often woolly growth on the surface of damp or decaying organic matter

moult¹, *NAm chiefly* **molt** *vb* to shed or cast off (hair, feathers, shell, horns, or an outer layer) periodically

moult², *NAm chiefly* **molt** *n* moulting; *specif* ecdysis

mouth *n, pl* **mouths** the opening through which food passes into an animal's body; *also* the cavity in the head of the typical vertebrate animal bounded externally by the lips that encloses the tongue, gums, and teeth

mouthpart *n* a structure or appendage near or forming part of the mouth

mRNA *n* MESSENGER RNA

mucilage *n* a gelatinous substance obtained esp from seaweeds and similar to plant gums

mucilaginous *adj* 2 of, full of, or secreting mucilage

mucosa *n, pl* **mucosae, mucosas** MUCOUS MEMBRANE

mucous *adj* of, like, secreting, or covered (as if) with mucus

mucous membrane *n* a membrane rich in mucous glands, specif lining body passages and cavities (eg the mouth) with openings to the exterior

mucronate *adj* having a sharp end, point, or part (eg of a base or apex of a leaf) ☞ PLANT

mucus *n* a thick slippery secretion produced by mucous membranes (eg in the nose) which it moistens and protects

mulch *n* a protective covering (eg of compost) spread on the ground to control weeds, enrich the soil, etc

mull *n* crumbly soil humus forming a layer of mixed organic matter and mineral soil and merging into the underlying mineral soil

Müllerian mimicry *adj* of or being mimicry between 2 or more inedible or dangerous species, considered to reduce the difficulties of recognition by potential predators

multicellular *adj* having or consisting of many cells

multilayered, multilayer *adj* having or involving several distinct layers, strata, or levels <~ *tropical rain forest*>

multinucleate *adj* having more than two nuclei

multiparous *adj* 1 producing many or more than 1 offspring at a birth 2 having given birth 1 or more times previously

multiple *adj* 1 consisting of or involving more than one 2 having numerous aspects or functions 3 formed by coalescence of the ripening ovaries or several flowers

multiple sclerosis *n* progressively developing partial or complete paralysis and jerking muscle tremor resulting from the formation of patches of hardened nerve tissue

mum

in nerves of the brain and spinal cord that have lost their myelin

mumps *n pl but sing or pl in constr* an infectious virus disease marked by gross swelling of esp the parotid glands

murmur *n* an atypical sound of the heart indicating an abnormality

muscle *n* (an organ that moves a body part, consisting of) a tissue made of modified elongated cells that contract when stimulated to produce motion

muscular *adj* **1** of, constituting, or performed by muscle or the muscles **2** having well-developed musculature

muscular dystrophy *n* progressive wasting of muscles occurring as a hereditary disease

musculature *n* the system of muscles of (part of) the body

mushroom *n* the enlarged, esp edible, fleshy fruiting body of a class of fungus, consisting typically of a stem bearing a flattened cap

mutable *adj* capable of or subject to mutation

mutagen *n* sthg (eg mustard gas) that increases the frequency of mutation

mutation *n* (an individual or strain differing from others of its type and resulting from) a relatively permanent change in an organism's hereditary material

mute¹ *adj* unable to speak; dumb

mute² *n* one who cannot or does not speak

mute³ *vi, of a bird* to pass waste matter from the body

mutualism *n* symbiosis for mutual benefit

mycelium *n, pl* **mycelia** the mass of interwoven filamentous hyphae that forms the body of a fungus and is usu submerged in another body (eg of soil or the tissues of a host)

mycetophagous *adj* feeding on fungi

mycetozoan *n* SLIME MOULD

mycoflora *n* the fungi characteristic of a region or special environment

mycology *n* (the biology of) fungal life or fungi

mycoplasma *n, pl* **mycoplasmas, mycoplasmata** any of a genus of minute microorganisms without cell walls that are intermediate in some respects between viruses and bacteria and are mostly parasitic, usu in mammals

mycorrhiza *n, pl* **mycorrhizae, mycorrhizas** the symbiotic association of the mycelium of a fungus with the roots of a flowering plant (eg an orchid)

mycosis *n, pl* **mycoses** infection with or disease caused by a fungus

mycotoxin *n* a toxic substance produced by a fungus, esp a mould

myelin *n* a soft white fatty material that forms a thick sheath around the cytoplasmic core of nerve cells adapted for fast conduction of nervous impulses

myelinated *adj, of a nerve fibre* having a sheath of myelin

myelitis *n* inflammation of the bone marrow

myelogenous, myelogenic *adj* of, originating in, or produced by the bone marrow

myeloid *adj* myelogenous

myeloma *n* a tumour of the bone marrow

myofibril *n* any of the long thin parallel contractile filaments of a muscle cell

myoglobin *n* a red iron-containing protein pigment in muscles, similar to haemoglobin

myopia *n* defective vision of distant objects resulting from the focussing of the visual images in front of the retina; shortsightedness — compare HYPERMETROPIA

myosin *n* a fibrous muscle protein that reacts with actin to produce muscular movement

myriapod, myriopod *n* a millipede, centipede, or related arthropod with a body made up of numerous similar segments bearing jointed legs

myrmecology *n* the study of ants

myxoedema *n* thickening and dryness of the skin and loss of vigour resulting from severe hypothyroidism

myxoma *n, pl* **myxomas, myxomata**

a soft tumour made up of gelatinous connective tissue

myxomatosis *n* a severe flea-transmitted virus disease of rabbits that is characterized by the formation of myxomas in the body, and that has been used in their biological control

myxomycete *n* SLIME MOULD

n *n, pl* **n's, ns** the haploid or gametic number of chromosomes

NAD *n* a widely occurring compound that is a cofactor of numerous enzymes that catalyse oxidation or reduction reactions [*ni*cotinamide-*a*denine *d*inucleotide]

nail *n* (a claw or other structure corresponding to) a horny sheath protecting the upper end of each finger and toe of human beings and other primates

naked *adj* **1** *of (part of) a plant or animal* lacking hairs or other covering or enveloping parts (eg a shell or feathers) **2** lacking foliage or vegetation

narcosis *n, pl* **narcoses** stupor or unconsciousness produced by narcotics or other chemicals

narcotic[1] *n* a substance or drug (eg opium) that soothes, relieves, or lulls

narcotic[2] *adj* **1a** like, being, or yielding a narcotic **b** inducing mental lethargy; soporific **2** of (addiction to) narcotics

naris *n, pl* **nares** the opening of the nose or nasal cavity of a vertebrate

nas

nasal *adj* of the nose

nastic *adj* of or being a movement of a plant part caused by the disproportionate growth of 1 surface

natal *adj* of, present at, or associated with (one's) birth

natality *n* the birthrate

natatorial, natatory *adj* (adapted to) swimming

native[1] *adj* **1** living (naturally), grown, or produced in a particular place; indigenous **2** *chiefly Austr* (superficially) resembling a specified British plant or animal

native[2] *n* a plant, animal, etc indigenous to a particular locality

natural history *n* **1** a treatise on some aspect of nature **2** the natural development of an organism, disease, etc over a period of time **3** the usu amateur study, esp in the field, of natural objects (eg plants and animals), often in a particular area

naturalist *n* a student of natural history

natural·ize, -ise *vt* to cause (eg a plant) to become established as if native ~ *vi* to become naturalized

natural science *n* any of the sciences (eg physics or biology) that deal with objectively measurable phenomena

natural selection *n* a natural process that tends to result in the survival of organisms best adapted to their environment and the elimi-

nation of (mutant) organisms carrying undesirable traits

nature *n* **1** a creative and controlling force in the universe **2** the inner forces in an individual **3** the physical constitution of an organism **4** the external world in its entirety

naturopathy *n* treatment of disease emphasizing stimulation of the natural healing processes, including the use of herbal medicines

nausea *n* a feeling of discomfort in the stomach accompanied by a distaste for food and an urge to vomit

navel *n* a depression in the middle of the abdomen marking the point of former attachment of the umbilical cord

navicular *n or adj* (a bone, esp in the ankle) shaped like a boat

Nearctic *adj* of or being the biogeographic subregion that includes Greenland and arctic and temperate N America

nearsighted *adj* able to see near things more clearly than distant ones; myopic

nebula *n, pl* **nebulas, nebulae** a cloudy patch on the cornea

necrophagous *adj* feeding on corpses or carrion < ~ *insects*>

necropsy *n* [2]POSTMORTEM

necrosis *n, pl* **necroses** (localized) death of living tissue

necrotic *adj* afflicted with necrosis

nectar *n* a sweet liquid secreted by

148

the flowers of many plants that is the chief raw material of honey

nectary *n* a plant gland that secretes nectar

need *n* a physiological or psychological requirement for the wellbeing of an organism

needle *n* 1 the slender hollow pointed end of a hypodermic syringe for injecting or removing material 2 a needle-shaped leaf, esp of a conifer

negative *adj* 1 not showing the presence or existence of the organism, condition, etc in question 2 directed or moving away from a source of stimulation <~ *tropism*>

nekton *n* aquatic animals (eg whales or squid) free-swimming near the surface of the water

nematocyst *n* any of the minute stinging organs of jellyfish or other coelenterates

nematode *n* any of a phylum of elongated cylindrical worms parasitic in animals or plants or free-living in soil or water

neoplasm *n* an abnormal growth of a tissue; a tumour

neoteny *n* 1 attainment of sexual maturity during the larval stage (eg in some salamanders) 2 retention of some larval or immature characters in adulthood

nephridium *n* a tubular glandular excretory organ characteristic of various coelomates

nephritic *adj* 1 renal 2 of or affected with nephritis

nephritis *n, pl* **nephritides** inflammation of the kidneys

nephron *n* a single excretory unit, esp of the kidneys of vertebrate animals

neritic *adj* of or being the region of shallow water adjoining the seacoast

nerve *n* 1 any of the filaments of nervous tissue that conduct nervous impulses to and from the nervous system and are made up of axons and dendrites 2 VEIN 3 3 the sensitive pulp of a tooth

nerve cell *n* a neuron

nerve gas *n* a deadly usu organophosphate poison gas that interferes with nerve transmission

nervous *adj* of, affected by, or composed of (the) nerves or neurons

nervous breakdown *n* (an occurrence of) a disorder in which worrying, depression, severe tiredness, etc prevent one from coping with one's responsibilities

nervous system *n* the brain, spinal cord, or other nerves and nervous tissue together forming a system for interpreting stimuli from the sense organs and transmitting impulses to muscles, glands, etc

neural *adj* 1 of or affecting a nerve or the nervous system 2 dorsal

neuralgia *n* intense paroxysms of

pain radiating along the course of a nerve without apparent cause

neuritis *n* inflammation or degeneration of a nerve causing pain, sensory disturbances, etc

neurochemistry *n* the biochemistry of (the transmission of impulses down) nerves

neuroglia *n* supporting tissue that is intermingled with the impulse-conducting cells of nervous tissue in the brain, spinal cord, and ganglia

neurology *n* the study of (diseases of) the nervous system

neuromuscular *adj* involving nervous and muscular cells, tissues, etc <*a ~ junction*>

neuron *n* any of the many specialized cells each with an axon and dendrites that form the functional impulse-transmitting units of the nervous system

neuropathy *n* an abnormal (degenerative) state of the nerves or nervous system

neuropteran *n* any of an order of insects, usu having a fine network of veins in their wings, including the lacewings

neurosis *n, pl* **neuroses** a nervous disorder, unaccompanied by disease of the nervous system, in which phobias, compulsions, anxiety, and obsessions make normal life difficult

neurotransmitter *n* a substance (eg acetylcholine) that is released at a nerve ending and transmits nerve impulses across the synapse

neuter[1] *adj* lacking generative organs or having nonfunctional ones <*the worker bee is ~*>

neuter[2] *n* a worker

neuter[3] *vt* to castrate

neutrophil, neutrophile *n* a white blood cell that has neutrophilic granules in its cytoplasm and is present in large numbers in the blood

neutrophilic, neutrophil *adj* staining weakly with both acidic and basic dyes

New World *n* the W hemisphere; *esp* the continental landmass of N and S America

niacin *n* NICOTINIC ACID

niche *n* the ecological role of an organism in a community, esp in regard to food consumption

nicotinamide *n* a vitamin of the vitamin B complex with actions similar to those of nicotinic acid

nicotinamide-adenine dinucleotide *n* NAD

nicotinic acid *n* a vitamin of the vitamin B complex that is found widely in animals and plants and whose lack results in pellagra

nictitating membrane *n* a thin membrane capable of extending across the eyeball under the eyelids of many animals (eg cats)

nidification *n* the act, process, or technique of building a nest

nidus *n, pl* **nidi, niduses** a nest or

breeding place; *esp* a place in an animal or plant where bacteria or other organisms lodge and multiply

night blindness *n* reduced vision in faint light (eg at night)

nipple *n* the small protuberance of a mammary gland (eg a breast) from which milk is drawn in the female

nit *n* (the egg of) a parasitic insect (eg a louse)

nitrification *n* nitrifying; *specif* the oxidation (eg by bacteria) of ammonium salts first to nitrites and then to nitrates

nitrogen *n* a trivalent gaseous chemical element that constitutes about 78 per cent by volume of the atmosphere and is found in combined form as a constituent of all living things

nitrogen cycle *n* the continuous circulation of nitrogen and nitrogen-containing compounds from air to soil to living organisms and back to air, involving nitrogen fixation, nitrification, decay, and denitrification ➢ ECOLOGY

nitrogen fixation *n* (industrial or biological) assimilation of atmospheric nitrogen into chemical compounds; *specif* this process performed by soil microorganisms, esp in the root nodules of leguminous plants (eg clover) ➢ ECOLOGY

nocturnal *adj* active at night <*a ~ predator*>

node *n* **1** a thickening or swelling (eg of a rheumatic joint) **2** a point on a stem at which 1 or more leaves are attached

node of Ranvier *n* a constriction in the myelin sheath of a myelinated nerve fibre

nodule *n* a small rounded mass: eg **a** a small rounded lump of a mineral or mineral aggregate **b** a swelling on the root of a leguminous plant (eg clover) containing symbiotic bacteria that convert atmospheric nitrogen into a form in which it can be used by the plant

nomenclature *n* a system of terms used in a particular science, discipline, or art

noradrenalin, noradrenalinen a compound from which adrenalin is formed in the body and which is the major neurotransmitter released from the nerve endings of the sympathetic nervous system

norm *n* the average: eg **a** a set standard of development or achievement, usu derived from the average achievement of a large group **b** a pattern typical of a social group

normal¹ *adj* **1** occurring naturally <*~ immunity*> **2** of, involving, or being a normal curve or normal distribution

normal² *n* sby or sthg that is normal

normal curve *n* the symmetrical bell-shaped curve of a normal distribution

normal distribution *n* a frequency distribution whose graph is a normal curve

nose¹ *n* **1a** the part of the face that bears the nostrils and covers the front part of the nasal cavity (together with the nasal cavity itself) **b** the front part of the head above or projecting beyond the mouth; a snout, muzzle **2** the sense or (vertebrate) organ of smell ➔ SENSE ORGAN

nose² *vt* to detect (as if) by smell; scent ~ *vi* to use the nose in examining, smelling, etc; to sniff or nuzzle

nostoc *n* any of a genus of blue-green algae able to fix nitrogen

nostril *n* the opening of the nose to the outside (together with the adjoining nasal passage)

notochord *n* a longitudinal rod that forms the supporting axis of the body in the lancelet, lamprey, etc and in the embryos of higher vertebrates

nourish *vt* to provide or sustain with nutriment; feed

nourishment *n* **1** food, nutriment **2** nourishing or being nourished

noxious *adj* harmful to living things <~ *industrial wastes*>

nuclear *adj* of or constituting a nucleus

nuclear membrane *n* the boundary of a cell nucleus

nuclease *n* any of various enzymes that promote the breakdown of nucleic acids

nucleate *vb* to form (into) a nucleus; to cluster

nucleated, nucleate *adj* having a nucleus or nuclei <~ *cells*>

nucleic acid *n* RNA, DNA, or another acid composed of a chain of nucleotide molecules linked to each other and found esp in cell nuclei

nuclein *n* a nucleoprotein

nucleolus *n, pl* **nucleoli** a spherical body in the nucleus of a cell that is prob the site of the synthesis of ribosomes

nucleoprotein *n* a compound of a protein (eg a histone) with a nucleic acid (eg DNA), forming the major constituent of chromosomes

nucleoside *n* any of several compounds (eg adenosine) consisting of a purine or pyrimidine base combined with deoxyribose or ribose and occurring esp as a constituent of nucleotides

nucleotide *n* any of several compounds that form the structural units of RNA and DNA and consist of a nucleoside combined with a phosphate group

nucleus *n, pl* **nuclei** *also* **nucleuses** a central point, mass, etc about which gathering, concentration, etc takes place: eg **a** a usu round membrane-surrounded cellular organelle containing the chromosomes **b** a (discrete) mass of nerve cells in the brain or spinal cord

null hypothesis *n* a statistical hypothesis to be tested and accepted or rejected in favour of an alternative

nulliparous *adj* not having borne offspring

nuptial *adj* characteristic of or occurring in the breeding season <*a ~ flight*>

nuptial plumage *n* the brilliantly coloured plumage developed in the males of many birds prior to the start of the breeding season — compare ECLIPSE PLUMAGE

nurture *n* 1 nourishment 2 all the environmental influences that affect the innate genetic potentialities of an organism

nut *n* (the often edible kernel of) a dry fruit or seed with a hard separable rind or shell

nutation *n* a spontaneous (spiral) movement of a growing plant part

nutrient *n or adj* (sthg) that provides nourishment

nutriment *n* sthg that nourishes or promotes growth

nutrition *n* nourishing or being nourished; *specif* all the processes by which an organism takes in and uses food

nutritious *adj* nourishing

nutritive *adj* 1 of nutrition 2 nourishing

nyctitropic *n* of or being a movement of a plant part at nightfall (eg the closing of a flower)

nymph *n* any of various immature insects; *esp* a larva of a dragonfly or other insect with incomplete metamorphosis ➔ LIFE CYCLE

nymphomania *n* excessive sexual desire in a female — compare SATYRIASIS

nystagmus *n* a rapid involuntary oscillation of the eyeballs (eg from dizziness)

oak apple *n* a large round gall produced on oak stems or leaves by a gall wasp

oasis *n, pl* **oases** a fertile or green area in a dry region

oat *n* 1a (any of various wild grasses related to) a widely cultivated cereal grass — usu pl **b** *pl* a crop or plot of oats 2 an oat seed

obligate *adj* 1 restricted to 1 characteristic mode of life <*an ~ parasite*> 2 always happening irrespective of environmental conditions <*~ parasitism*> — compare FACULTATIVE

oblique *adj, of a muscle* situated obliquely with 1 end not attached to bone

obovate *adj, of a leaf* ovate with the narrower end nearest the stalk

obsession *n* a persistent (disturbing) preoccupation with an often unreasonable idea; *also* an idea causing such a preoccupation

obtuse *adj, of a leaf* rounded at the end furthest from the stalk ➔ PLANT

obverse *adj* with the base narrower than the top <*an ~ leaf*>

occipital *adj* of, situated near, or being the back part of the head or skull

occiput *n, pl* **occiputs, occipita** the back part of the head

occult *adj* not present, manifest, or detectable by the unaided eye <*~ blood loss*>

occupational therapy *n* creative activity used as therapy for promoting recovery or rehabilitation

oceanarium *n, pl* **oceanariums, oceanaria** a large marine aquarium

oceanic *adj* of, produced by, or occurring in the ocean, esp the open sea

ocellus *n, pl* **ocelli** a minute simple eye or eyespot of an invertebrate animal (eg an insect) ☞ INSECT

octopod *n* any of an order of cephalopod molluscs with 8 arms bearing sessile suckers, including the octopus

ocular *adj* of the eye <*~ muscles*>

odontoid process *n* a toothlike projection from the front end of the second vertebra in the neck on which the first vertebra and the head rotate

odour, *NAm chiefly* **odor** *n* (the sensation resulting from) a quality of sthg that stimulates the sense of smell

oedema, *NAm chiefly* **edema** *n* abnormal accumulation of liquid derived from serum causing abnormal swelling of the tissues

Oedipus complex *n* (an adult personality disorder resulting from) the sexual attraction developed by a child towards the parent of the opposite sex with accompanying jealousy of the parent of the same sex

oesophagus *n, pl* **oesophagi** the muscular tube leading from the back of the mouth to the stomach

oestrogen *n* a substance, esp a sex hormone, that stimulates the development of secondary sex characteristics in female vertebrates and promotes oestrus in lower mammals

oestrus *n* a regularly recurrent state of sexual excitability in the female of most lower mammals when she will copulate with the male

oestrus cycle *n* the series of changes in a female mammal occurring from one period of oestrus to the next

offspring *n, pl* **offspring** the progeny of a person, animal, or plant; young

oidium *n, pl* **oidia 1** any of (the small asexual spores borne in chains by) various fungi many of which are the spore-bearing stages of powdery mildews **2** a powdery mildew, esp on grapes, caused by an oidium

Old World *n* the E Hemisphere; *specif* Europe, Asia, and Africa

olfaction *n* smelling or the sense of smell

olfactory organ *n* a (membranous) organ of the sense of smell (situated in the nasal cavity)

Oligocene *adj* of or being an epoch of the Tertiary between the Eocene and Miocene

oligochaete *n or adj* (any) of a class of freshwater and ground-living annelid worms (eg the earthworm) with relatively few bristles along the body — compare POLYCHAETE

oligotrophic *adj* deficient in plant nutrients; *esp* of a body of water having abundant dissolved oxygen and no marked stratification —compare EUTROPHIC

omasum *n, pl* **omasa** the third stomach of a ruminant mammal, lying between the reticulum and the abomasum

omnivorous *adj* feeding on both animal and vegetable substances

oncology *n* the study and treatment of cancer and malignant tumours

ontogeny *n* the (course of) development of an individual organism

oocyst *n* a zygote; *esp* a sporozoan zygote

oocyte *n* an egg before maturation or division to form female gametes

oogamete *n* a relatively large immobile female gamete

oogamous *adj* having or involving a small mobile male gamete and a large immobile female gamete

oogenesis *n* the formation and maturation of eggs or ova

oogonium *n* the female sexual organ in various algae and fungi

oosperm *n* a zygote

oospore *n* a fertilized plant spore that grows into the phase of a plant producing sexual spores — compare ZYGOSPORE

ootheca *n* a firm-walled and distinctive egg case (eg of a cockroach)

ooze *n* **1** a soft deposit of mud, slime, debris, etc on the bottom of a body of water **2** (the muddy ground of) a marsh or bog

open-heart *adj* of or performed on a heart surgically opened whilst its function is temporarily taken over by a heart-lung machine <~ *surgery*>

operation *n* a procedure carried out on a living body with special instruments, usu for the repair of damage or the restoration of health

operator gene *n* a region of the chromosome that triggers a structural gene to form mRNA and that is inhibited by a repressor

operculum *n, pl* **opercula** *also* **operculums 1** a lid or covering flap (eg of a moss capsule or the gills of a fish) **2** a hard plate at the end of the foot in many gastropod molluscs that closes the shell when the animal is retracted

operon *n* a set of genes on a

chromosome that function together as a unit

ophthalmia *n* inflammation of the conjunctiva or the eyeball

ophthalmic *adj* of or situated near the eye

ophthalmology *n* the branch of medical science dealing with the structure, functions, and diseases of the eye

ophthalmoscope *n* an instrument used to view the retina and other structures inside the eye

opposable *adj, of a thumb or other digit* capable of being placed opposite and against 1 or more of the remaining digits

opposite *adj, of plant parts* situated in pairs at the same level on opposite sides of an axis <~ leaves> — compare ALTERNATE ☞ PLANT

optic¹ *adj* of vision or the eye

optic² *n* the eye

optical *adj* visual <an ~ illusion>

optimum *n, pl* **optima** *also* **optimums** (the amount or degree of) sthg that is most favourable to a particular end

oral *adj* 1 of, given through, or affecting the mouth <~ contraceptive> 2 of or characterized by (passive dependency, aggressiveness, or other personality traits typical of) the first stage of sexual development in which gratification is derived from eating, sucking,

and later by biting — compare ANAL, GENITAL

orbicular *adj* 1 spherical 2 circular <~ leaves> ☞ PLANT

orbit *n* the bony socket of the eye

orchidectomy *n* the surgical removal of 1 or both testicles

orchitis *n* inflammation of the testicles

order *n* a category in the classification of living things ranking above the family and below the class

Ordovician *adj or n* (of or being) the period of the Palaeozoic era between the Cambrian and the Silurian

organ *n* a differentiated structure (eg the heart or a leaf) consisting of cells and tissues and performing some specific function in an organism

organelle *n* a part of a cell (eg a mitochondrion) that has a specialized structure and usu a specific function

organic *adj* 1a of or arising in a bodily organ b affecting the structure of the organism <an ~ disease> — compare FUNCTIONAL 1 2a of or derived from living organisms b of or being food produced using fertilizer solely of plant or animal origin without the aid of chemical fertilizers, pesticides, etc <~ farming>

organism *n* a living being

orgasm *n* intense or paroxysmal emotional excitement; *esp* (an instance of) the climax of sexual excitement, occurring typically as the culmination of sexual intercourse

orientation *n* change of position by (a part of) an organism in response to an external stimulus

orifice *n* an opening (eg a vent or mouth) through which sthg may pass

origin *n* the more fixed, central, or large attachment or part of a muscle

ornithology *n* a branch of zoology dealing with birds

orthopteran, orthopteron *n, pl* **orthopterans, orthoptera** any of an order of large insects (eg crickets and grasshoppers) with biting mouthparts and either no wings or 2 pairs of wings

osmoregulation *n* the usu automatic regulation of osmotic pressure, esp in the body of an organism

osmosis *n* movement of a solvent through a semipermeable membrane (eg of a living cell) into a solution of higher concentration that tends to equalize the concentrations on the 2 sides of the membrane

osmotic pressure *n* the pressure produced by or associated with osmosis and dependent on concentration and temperature

ossein *n* the collagen of bones

ossicle *n* a small bone or bony structure (eg in the middle ear)

ossify *vi* to become bone ∼ *vt* to change (eg cartilage) into bone

osteoarthritis *n* degenerative arthritis

osteomalacia *n* softening of the bones, esp in elderly people, equivalent to rickets in young people

osteomyelitis *n* an infectious inflammatory disease of bone (marrow)

osteopathy *n* a system of treatment of diseases based on the theory that they can be cured by manipulation of bones

osteophyte *n* an abnormal outgrowth from a bone

otic *adj* of or located in the region of the ear

otolith *n* any of many minute lumps of calcite and protein in the internal ear that are receptors for the sense of balance

outbreeding *n* interbreeding of relatively unrelated animals or plants

outer ear *n* the outer visible part of the ear together with the canal through which sound waves reach the eardrum

outgrowth *n* a process or product of growing out <an ∼ of hair>

ovariectomy *n* the surgical removal of an ovary

ovary *n* **1** a typically paired female reproductive organ that produces eggs and female sex hormones ☞ REPRODUCTION **2** the enlarged rounded usu basal female part of a flowering plant that bears the ovules and consists of 1 or more carpels ☞ FLOWER

ovate *adj* (having an outline) shaped like (a longitudinal section of) an egg <*an ~ leaf*> ☞ PLANT

overblown *adj* past the prime of bloom <*~ roses*>

overgraze *vt* to allow animals to graze to the point of damaging vegetational cover

overpopulation *n* the condition of having too dense a population, so that the quality of life is impaired

overwinter *vi* to survive or spend the winter

ovicide *n* an agent that kills eggs; *esp* an insecticide effective against the egg stage

oviduct *n* the tube that serves for the passage of eggs from an ovary, esp before laying

ovine *adj* of or resembling sheep

oviparous *adj* involving or producing eggs that develop and hatch outside the mother's body

oviposit *vi, esp of an insect* to lay eggs

ovipositor *n* a specialized organ, esp of an insect, for depositing eggs

ovoid *adj* egg-shaped

ovotestis *n* a hermaphrodite gonad (eg in some snails)

ovoviviparous *adj* producing eggs that develop and usu hatch within the mother's body

ovulate *vi* to produce eggs or discharge them from an ovary

ovule *n* an outgrowth of the ovary of a seed plant that develops into a seed after fertilization of the egg cell it contains ☞ FLOWER

ovum *n, pl* **ova** an animal's female gamete that when fertilized can develop into a new individual

oxidation-reduction *n* a chemical reaction in which one or more electrons are transferred from one atom or molecule to another

oxidative phosphorylation *n* the synthesis in mitochondria of ATP from ADP using energy obtained from the oxidation of substances formed during the Krebs cycle

oxygen *n* a bivalent gaseous chemical element that forms about 21 per cent by volume of the atmosphere, is found combined in water, most minerals, and many organic compounds, is required for most burning processes, and is essential for the life of all plants and animals

oxygenate *vt* to impregnate, combine, or supply (eg blood) with oxygen

oxygen debt *n* a cumulative oxygen lack that develops during intense activity and must be made up when the body returns to rest

oxyhaemoglobin *n* haemoglobin loosely combined with oxygen that it releases to the tissues

oxytocin *n* a polypeptide hormone

secreted by the back lobe of the pituitary gland that stimulates the contraction of uterine muscle (eg during childbirth) and the ejection of milk

pacemaker *n* (a device for applying regular electric shocks to the heart that reproduces the function of) a part of the heart that maintains rhythmic (coordinated) contractions

pachyderm *n* an elephant, rhinoceros, pig, or other usu thick-skinned (hoofed) nonruminant mammal

pachytene *n* the stage of the prophase of meiotic cell division in which the paired chromosomes become thickened and divided into chromatids

paediatrics *n* medicine dealing with the development, care, and diseases of children

pain¹ *n* **1** a basic bodily sensation induced by a noxious stimulus or physical disorder and characterized by physical discomfort (eg pricking, throbbing, or aching) **2** *pl* the throes of childbirth

pain² *vt* to make suffer or cause distress to; hurt ~ *vi* to give or have a sensation of pain

painful *adj* **-ll-** feeling or giving pain

Palaearctic *adj* of or being a biogeographic region that includes Europe and N Asia, Arabia, and Africa

palaeobotany *n* a branch of botany dealing with fossil plants

palaeoclimatology *n* the study of the climate of past ages

Palaeolithic *adj or n* (of or being) the 2nd era of the Stone Age characterized by rough or chipped stone implements

palaeontology *n* a science dealing with the life of past geological periods as inferred from fossil remains

palaeozoology *n* a branch of zoology dealing with fossil animals

Palaeozoic *adj* of or being the era of geological history from the beginning of the Cambrian to the end of the Permian

palate *n* **1** the roof of the mouth, separating it from the nasal cavity **2** the sense of taste

palatine *adj* of or lying near the palate

palea *n, pl* **paleae** a chaffy scale or bract; *esp* the upper bract of the flower of a grass

palingenetic *adj* of or being biological features (eg the gill slits in a human embryo) that are derivations from distant ancestral forms

palisade layer *n* a layer of cells containing many chloroplasts lying beneath the upper skin of green leaves

pallor *n* deficiency of (facial) colour; paleness

palmate *also* **palmated** *adj* (having lobes or veins radiating from a

pal

common point) resembling a hand with the fingers spread ⇒ PLANT

palp *n* a segmented (touch- or taste-sensitive) feeler on the mouthparts of an insect or other arthropod ⇒ INSECT

palpebral *adj* of or near the eyelids

palpitate *vi* to beat rapidly and strongly; throb <*a palpitating heart*>

palsy *n* paralysis or uncontrollable tremor of (a part of) the body

paludal *adj* of marshes or fens

palynology *n* a branch of botany dealing with pollen and spores

pampa *n* an extensive (grass-covered) plain of temperate S America east of the Andes — usu pl with sing. meaning but sing. or pl in constr

pan *n* **1** a natural basin or depression in land **2** hardpan

pancreas *n* a large compound gland in vertebrates that secretes digestive enzymes into the intestines and the hormones insulin and glucagon into the blood

pancreatic juice *n* the secretion of pancreatic digestive enzymes that is poured into the duodenum

pancreatin *n* (a preparation containing) a mixture of enzymes from the pancreatic juice

pandemic *n or adj* (a disease) occurring over a wide area and affecting an exceptionally high proportion of the population

panicle *n* a (pyramidal) loosely

branched flower cluster or compound inflorescence ⇒ FLOWER

panleucopenia *n* an acute usu fatal epidemic virus disease of cats

pantothenic acid *n* a vitamin of the vitamin B complex

papilionaceous *adj* having an irregular butterfly-shaped corolla

papilla *n, pl* **papillae** a small projecting nipple-shaped body part: eg **a** a piece of connective tissue extending into and nourishing the root of a hair, feather, etc **b** any of the protuberances of the dermal layer of the skin extending into the epidermal layer **c** any of the protuberances on the upper surface of the tongue

papilloma *n, pl* **papillomas, papillomata** a benign tumour (eg a wart) due to overgrowth of epithelial tissue

pappus *n, pl* **pappi** a (tuft of) usu hairy appendages crowning the ovary or fruit in various plants (eg the dandelion)

papule *n* a small solid usu conical projection from the skin

parabiosis *n* anatomical and physiological union of 2 organisms

paradoxical sleep *n* a state of sleep that is characterized esp by dreaming, rapid eye movements, and vascular congestion of the sex organs

paralysis *n, pl* **paralyses** **1** (partial) loss of function, esp when involving motion or sensation in a part of the body **2** loss of the ability to move

paranoia *n* **1** a mental disorder characterized by delusions of persecution or grandeur **2** a tendency towards excessive or irrational suspiciousness and distrustfulness of others

paraplegia *n* paralysis of the lower half of the body including the legs

parapsychology *n* the investigation of evidence for the occurrence of psychic phenomena (eg telepathy and clairvoyance)

parasite *n* an organism living in or on another organism in parasitism

parasitism *n* an intimate association between organisms of 2 or more kinds in which a parasite benefits at the expense of a host

parasympathetic *adj* of, being, mediated by, or acting on (the nerves of) the parasympathetic nervous system

parasympathetic nervous system *n* the part of the autonomic nervous system that contains nerve fibres in which the neurotransmitter is acetylcholine and whose activity tends to contract smooth muscle and cause the dilation of blood vessels — compare SYMPATHETIC NERVOUS SYSTEM

parathyroid, parathyroid gland *n* any of 4 small endocrine glands near the thyroid gland that produce a hormone

paratyphoid *n* a disease caused by

salmonella that resembles typhoid fever and is commonly contracted by eating contaminated food

parenchyma *n* **1** a fleshy tissue of the leaves, fruits, stems, etc of higher plants that consists of thin-walled living cells — compare COLLENCHYMA, SCLERENCHYMA **2** the essential and distinctive tissue of an organ or an abnormal growth, as distinguished from its supportive framework

parent *n* **1** sby who begets or brings forth offspring; a father or mother **2** an animal or plant regarded in relation to its offspring

parenteral *adj* situated, occurring, or administered outside the intestines

paresis *n, pl* **pareses** slight or partial paralysis

parietal *adj* **1** of the walls of an anatomical part or cavity **2** of or forming the upper rear wall of the skull

parietal bone *n* either of a pair of bones of the top and side of the skull

Parkinson's disease *n* tremor, weakness of resting muscles, and a peculiar gait occurring in later life as a progressive nervous disease

parotid gland *n* either of a pair of large salivary glands below and in front of the ear

paroxysm *n* a fit, attack, or sudden increase or recurrence of (disease)

par

symptoms; a convulsion <*a ~ of coughing*>

part *n* an organ, member, or other constituent element of a plant or animal body

parthenogenesis *n* reproduction by development of an unfertilized gamete that occurs esp among lower plants and invertebrate animals

partite *adj* cleft nearly to the base <*a ~ leaf*>

parturition *n* the action or process of giving birth to offspring

passage *n* **1** passing sthg or undergoing a passing **2** incubation of a pathogen (eg a virus) in culture, a living organism, or a developing egg

passerine *adj* of the largest order of birds that consists chiefly of perching songbirds (eg finches, warblers, and thrushes)

passive *adj* **1** *of a person* lacking in energy, will, or initiative; meekly accepting **2** not involving expenditure of chemical energy <*~ transport across a cell membrane*>

pastern *n* (a part of an animal's leg corresponding to) a part of a horse's foot extending from the fetlock to the hoof

pasteur·ization, -isation *n* partial sterilization of a substance, esp a liquid (eg milk), by heating for a short period

pasturage *n* pasture

pasture *n* **1** plants (eg grass) grown for feeding (grazing) animals **2** (a plot of) land used for grazing **3** the feeding of livestock; grazing

patagium *n, pl* **patagia** a wing membrane; *esp* the fold of skin connecting the forelimbs and hind limbs of a gliding animal (eg a flying squirrel)

patella *n, pl* **patellae, patellas** the kneecap ➔ ANATOMY

paternal *adj* received or inherited from one's male parent

paternity test *n* the comparison of the genetic attributes (eg blood groups) of a mother, child, and man to determine whether the man could be the child's father

pathogen *n* a bacterium, virus, or other disease-causing agent

pathogenesis *n* the origination and development of a disease

pathogenic *adj* (capable of) causing disease

pathologist *n* one who studies pathology; *specif* one who conducts postmortems to determine the cause of death

pathology *n* **1** the study of (the structure and functional changes produced by) diseases **2** sthg abnormal: **a** the anatomical and physiological abnormalities that constitute or characterize (a particular) disease **b** deviation from an assumed normal state of mentality or morality

pathway *n* the sequence of enzyme-catalysed reactions by which a substance is synthesized or an energy-yielding substance is used by living tissue <*metabolic* ~s>

paw *n* the (clawed) foot of a lion, dog, or other (quadruped) animal

peat *n* (a piece of) partially carbonized vegetable tissue formed by partial decomposition in water of various plants (eg mosses), found in large bogs, and used esp as a fuel for domestic heating and as a fertilizer

peat moss *n* sphagnum

pecking order, peck order *n* 1 the natural hierarchy within a flock of birds, esp poultry, in which each bird pecks another lower in the scale without fear of retaliation 2 a social hierarchy

pectin *n* any of various water-soluble substances that bind adjacent cell walls in plant tissues and yield a gel which acts as a setting agent in jams and fruit jellies

pectinate, pectinated *adj* having narrow parallel projections or divisions suggestive of the teeth of a comb <~ *antennae*>

pectoral *adj* of, situated in or on, or worn on the chest

pectoral fin *n* either of the fins of a fish that correspond to the forelimbs of a quadruped

pectoral girdle *n* the bony or cartilaginous arch that supports the forelimbs of a vertebrate

ped *n* a natural soil aggregate

pedicel *n* 1 a plant stalk that sup-

ports a fruiting or spore-bearing organ 2 a narrow basal attachment of an animal organ or part

pedicle *n* a pedicel

pediculosis *n* infestation with lice

pedicure *n* 1 one who practises chiropody 2 (a) treatment for the care of the feet and toenails

pedigree[1] *n* the recorded purity of breed of an individual or strain

pedigree[2] *adj* of, being, or producing pedigree animals

pedipalp *n* either of the second pair of appendages of an arachnid (eg a spider) that are near the mouth and are often modified for a special (eg sensory) function

pedology *n* SOIL SCIENCE

peduncle *n* 1 a stalk bearing a flower, flower cluster, or fruit 2 a narrow stalklike part by which some larger part or the whole body of an organism is attached

peel *n* the skin or rind of a fruit

pelage *n* the hairy covering of a mammal

pelagic *adj* of, occurring, or living (at or above moderate depths) in the open sea — compare DEMERSAL

pellagra *n* dermatitis and nervous symptoms associated with a deficiency of nicotinic acid and protein in the diet

pellicle *n* a thin skin or film

peltate *adj* shaped like a shield; *specif, of a leaf* having the stem or support attached to the lower surface instead of at the base or margin ⊃ PLANT

pelvic girdle *n* the bony or cartilaginous arch that supports the hind limbs of a vertebrate

pelvis *n, pl* **pelvises, pelves** **1** (the cavity of) a basin-shaped structure in the skeleton of many vertebrates that is formed by the pelvic girdle and adjoining bones of the spine **2** the funnel-shaped cavity of the kidney into which urine is discharged ⊃ ANATOMY

penicillate *adj* having a tuft of fine filaments <*a ~ stigma*>

penicillin *n* (a salt, ester, or mixture of salts and esters of) any of several antibiotics or antibacterial drugs orig obtained from moulds, that act by interfering with the synthesis of bacterial cell walls and are active against a wide range of bacteria

penicillium *n* any of a genus of fungus found as a blue mould on decaying organic matter; *esp* a fungus producing penicillin

penile *adj* of or affecting the penis

penis *n, pl* **penes, penises** the male organ of copulation by which semen is introduced into the female during coitus ⊃ REPRODUCTION

pentamerous *adj* divided into or consisting of 5 parts; *specif, of a flower* having each whorl of petals, sepals, stamens, etc consisting of (a multiple of) 5 members

pentose *n* any of various monosac-charide sugars (eg ribose) that contain 5 carbon atoms in the molecule

pepo *n, pl* **pepos** a fleshy many-seeded berry (eg a pumpkin, melon, or cucumber) with a hard rind

pepsin *n* an enzyme of the stomach that breaks down most proteins in an acid environment

peptic *adj* **1** of or promoting digestion **2** connected with or resulting from the action of digestive juices <*a ~ ulcer*>

peptide *n* a short chain of 2 or more amino acids joined by peptide bonds

peptide bond *n* the chemical bond between the carbon of one amino acid and the nitrogen of another that links amino acids in peptides and proteins

peptone *n* any of various water-soluble products of protein breakdown

percentile *n* a statistical measure (eg used in educational and psychological testing) that expresses a value as a percentage of all the values that are lower than or equal to it

perception *n* **1a** a result of perceiving; an observation **b** a mental image; a concept **2** the mental interpretation of physical sensations produced by stimuli from the external world

perennial *adj, of a plant* living for

164

several years, usu with new herbaceous growth each year

perfect *adj, of a plant* having the stamens and carpels in the same flower

perfuse *vt* to force a fluid through (an organ or tissue), esp by way of the blood vessels

perianth *n* the external envelope of a flower, esp when not differentiated into petals and sepals

pericardium *n, pl* **pericardia** the membranous sac that surrounds the heart of vertebrates

pericarp *n* the ripened wall of a plant ovary

pericranium *n, pl* **pericrania** the external membrane of the skull

pericycle *n* a thin layer of cells that surrounds the central vascular part of many stems and roots

periderm *n* a thick outer protective tissue layer of woody roots and stems that consists of cork and adjacent tissues

perigynous *adj* (having floral organs) borne on a ring or cup of the receptacle surrounding an ovule — compare EPIGYNOUS, HYPOGYNOUS

perilymph *n* the liquid inside the labyrinth of the inner ear

perinatal *adj* (occurring) at about the time of birth

perineum *n* the area between the anus and the back part of the genitals, esp in the female

periodicity *n* the state or fact of being regularly recurrent

periodontal *adj* (of or affecting tissues) surrounding a tooth

periosteum *n, pl* **periostea** the membrane of connective tissue that closely surrounds all bones except at the joints

peripatus *n* any of a class of primitive tropical arthropods that in some respects are intermediate between annelid worms and typical arthropods

peripheral *adj* **1** located away from the centre **2** of, using, or being the outer part of the field of vision <*good* ~ *vision*>

periphery *n* the external boundary or surface of a (person's) body, esp as distinguished from its internal regions or centre

peristalsis *n* successive waves of involuntary contraction passing along the walls of a hollow muscular structure, esp the intestine, and forcing the contents onwards

peritoneum *n, pl* **peritoneums, peritonea** the smooth transparent membrane that lines the cavity of the mammalian abdomen

peritonitis *n* inflammation of the peritoneum

permeable *adj* capable of being permeated; *esp* having pores or openings that permit liquids or gases to pass through <*a* ~ *membrane*>

Per

Permian *adj or n* (of or being) the last period of the Palaeozoic era

pernicious anaemia *n* anaemia marked by a decrease in the number of red blood cells which is caused by a reduced ability to absorb vitamin B$_{12}$

perpetual *adj, of a plant* blooming continuously throughout the season

persistent *adj* 1 continuing to exist in spite of interference or treatment <a ~ cough> 2a remaining (1) beyond the usual period <a ~ leaf> (2) without change in function or structure <~ gills> b of a chemical substance broken down only slowly in the environment <~ pesticides>

persona *n, pl* **personas** an individual's social facade that, esp in Jungian psychology, reflects the role that the individual is playing in life — compare ANIMA

personality *n* the totality of an individual's behavioural and emotional tendencies; broadly a distinguishing complex of individual or group characteristics

perspiration *n* 1 sweating 2 ²SWEAT

perspire *vi* ¹SWEAT 1

pertussis *n* WHOOPING COUGH

pest *n* 1 a pestilence 2 a plant or animal capable of causing damage or carrying disease

pesticide *n* a chemical used to destroy insects and other pests of crops, domestic animals, etc

pestilence *n* a virulent and devastating epidemic disease; *specif* BUBONIC PLAGUE

pestilent *adj* destructive of life; deadly

petal *n* any of the modified often brightly coloured leaves of the corolla of a flower ⟹ FLOWER

petalous *adj* having (such or so many) petals — usu in combination <*poly*petalous>

petiole *n* the usu slender stalk by which a leaf is attached to a stem

petri dish *n* a small shallow glass or plastic dish with a loose cover used esp for cultures of microorganisms (eg bacteria)

petrifaction *n* 1 the process of petrifying; being petrified 2 sthg petrified

petrify *vt* to convert (as if) into stone or a stony substance

petrous *adj* resembling stone, esp in hardness; *specif* of or being the hard dense part of the human temporal bone that contains the internal hearing organs

phage *n* a bacteriophage

phagocyte *n* a macrophage, white blood cell, etc that characteristically engulfs foreign material (eg bacteria) and consumes debris (eg from tissue injury)

phagocytosis *n, pl* **phagocytoses** the uptake and usu destruction of extracellular solid matter by phagocytes — compare PINOCYTOSIS

phalanx *n, pl* **phalanges, phalanxes**

any of the digital bones of the hand or foot of a vertebrate ⮕ ANATOMY

phanerophyte *n* a perennial plant that bears its overwintering buds well above the surface of the ground

pharmaceutics *n pl but sing in constr* the science of preparing, using, or dispensing medicines

pharynx *n, pl* **pharynges** *also* **pharynxes** the part of the vertebrate alimentary canal between the mouth cavity and the oesophagus

phase-contrast microscope *n* a microscope that changes differences in the phase of the light transmitted through or reflected by the object into differences of intensity in the image and is used esp for examining biological specimens that have not been stained

phellem *n* an outer layer of cork cells produced in the roots or stems of woody plants by phellogen

phelloderm *n* a layer of (parenchyma) cells produced inwardly in the roots or stems of woody plants by phellogen

phellogen *n* a single row of cells in the outer layer of a woody plant stem or root that divides to form phellem to the outside and phelloderm to the inside

phenology *n* a branch of science dealing with relations between climate and periodic biological phenomena (eg bird migration or plant flowering)

phenotype *n* the visible characteristics of an organism that are produced by the interaction of the organism's genes and the environment

phenylalanine *n* an amino acid found in most proteins that is essential for human metabolism

pheromone *n* a chemical substance that is produced by an animal and stimulates 1 or more behavioural responses in other individuals of the same species

phlebitis *n* inflammation of a vein

phlebotomy *n* the letting or taking of blood in the treatment or diagnosis of disease

phlegm *n* thick mucus secreted in abnormal quantities in the respiratory passages

phloem *n* a complex vascular tissue of higher plants that functions chiefly in the conduction of soluble food substances (eg sugars) — compare XYLEM

phobia *n* an exaggerated and illogical fear of sthg

phobic *adj* 1 of or being a phobia 2 motivated by or based on withdrawal from an unpleasant stimulus <*a ~ response to light*>

phosphene *n* an impression of light due to excitation of the retina caused by pressure on the eyeball

phosphorylation *n* the combining of an organic compound with an

pho

inorganic phosphate group; *esp* the conversion of carbohydrates (eg glucose) into their phosphates in metabolic processes

photic *adj* 1 of or involving light, esp in its effect on living organisms 2 penetrated by (the sun's) light <~ *zone of the ocean*>

photochemistry *n* (chemistry that deals with) the effect of radiant energy in producing chemical changes

photogenic *adj* producing or generating light; luminescent <~ *bacteria*>

photoperiod *n* the relative lengths of alternating periods of lightness and darkness as they affect the growth and maturity of an organism

photophore *n* a light-emitting organ; *esp* any of the luminous spots on various marine mostly deep-sea fishes

photophosphorylation *n* the synthesis of ATP from ADP and phosphate that occurs in a plant using radiant energy absorbed during photosynthesis

photopic *adj* of or being vision in bright light with light-adapted eyes — compare SCOTOPIC

photoreceptor *n* a receptor for light stimuli

photosensitive *adj* sensitive or sensitized to radiant energy, esp light

photosensit·ize, **-ise** *vt* to make

(abnormally) sensitive to the influence of radiant energy, esp light

photosynthesis *n* the synthesis of organic chemical compounds from carbon dioxide and water using radiant energy, esp light; *esp* the formation of carbohydrates in the chlorophyll-containing tissues of plants exposed to light ☞ PLANT

phototropism *n* a tropism in which light is the orienting factor

phrenology *n* the study of the conformation of the skull as an indication of mental faculties and character

phycocyanin *n* any of various bluish green pigments in the cells of blue-green algae

phycoerythrin *n* any of the red pigments in the cells of red algae

phycomycete *n* any of a class of lower fungi similar to algae

phylloclade *n* a flattened stem or branch that functions as a leaf

phyllode *n* a flat expanded leaf stalk that resembles the blade of a foliage leaf and fulfils the same functions

phyllotaxy *also* **phyllotaxis** *n* (the study of) the arrangement of leaves on a stem

phylogenesis *n* phylogeny

phylogeny *n* (the history of) the evolution of a genetically related group of organisms (eg a race or species)

phylum *n, pl* **phyla** a major group of

168

pin

related species in the classification of plants and animals

physiological, physiologic *adj* **1** of physiology **2** characteristic of or appropriate to an organism's healthy or normal functioning <*the ~ level of a substance in the blood*>

physiology *n* **1** biology that deals with the functions and activities of life or of living matter (eg organs, tissues, or cells) and the physical and chemical phenomena involved — compare ANATOMY **2** the physiological activities of (part of) an organism or a particular bodily function <*the ~ of sex*>

physiotherapy *n* the treatment of disease by physical and mechanical means (eg massage and regulated exercise)

phytography *n* descriptive botany, sometimes including plant taxonomy

phytophagous *adj, esp of an insect* feeding on plants

phytoplankton *n* planktonic plant life — compare ZOOPLANKTON

phytotoxic *adj* poisonous to plants

pia mater *n* the thin membrane that envelops the brain and spinal cord and is internal to the dura mater

pictograph, pictogram *n* a diagram representing statistical data by pictorial forms

pigment *n* (a colourless substance related to) any of various colouring matters in animals and plants

pigmentation *n* (excessive) colora-

tion with, or deposition of, (bodily) pigment

pileus *n, pl* **pilei** the (umbrella-shaped) fruiting body of many fungi (eg mushrooms)

pilose *adj* covered with (soft) hair

pilus *n, pl* **pili** (a structure resembling) a hair

pincer *n* a claw (eg of a lobster) resembling a pair of pincers

pinch *vt* to prune the tip of (a plant or shoot), usu to induce branching — + *out* or *back*

pineal body *n* PINEAL GLAND

pineal gland *n* a small appendage of the brain of most vertebrates that has the structure of an eye in a few reptiles, and that secretes melatonin and other hormones

pinetum *n, pl* **pineta** a plantation of pine trees; *also* a scientific collection of living coniferous trees

pinion[1] *n* **1** (the end section of) a bird's wing **2** a bird's feather; a quill

pinion[2] *vt* to restrain (a bird) from flight, esp by cutting off the pinion of a wing

pinna *n, pl* **pinnae, pinnas 1** a leaflet or primary division of a pinnate leaf or frond **2** the largely cartilaginous projecting portion of the outer ear ⭢ SENSE ORGAN

pinnate *adj* resembling a feather, esp in having similar parts arranged on opposite sides of an axis like the barbs on the shaft of a feather <*a ~ leaf*> ⭢ PLANT

pin

pinnule *n* **1** any of the secondary branches of a pinnate leaf or organ **2** a small fish fin separated from a major fin

pinocytosis *n*, *pl* **pinocytoses** the uptake of extracellular fluid by a cell by invagination of the cell membrane and formation of a fluid-filled sac inside the cell — compare PHAGOCYTOSIS

piscine *adj* (characteristic) of fish

piscivorous *adj* feeding on fishes

pisiform *adj* pea-shaped

pistil *n* a carpel

pistillate *adj* having pistils but no stamens

pit *n* a hollow or indentation, esp in the surface of a living plant or animal: eg **a** a natural hollow in the surface of the body **b** any of the indented scars left in the skin by a pustular disease (eg smallpox)

pitcher *n* a modified leaf of a pitcher plant in which the hollowed stalk and base of the blade form an elongated receptacle

pith[1] *n* **1** a (continuous) central area of spongy tissue in the stems of most vascular plants **2** the white tissue surrounding the flesh and directly below the skin of a citrus fruit

pith[2] *vt* **1** to destroy the spinal cord or central nervous system of (eg cattle or a frog) **2** to remove the pith from (a plant part)

pituitary *adj or n* (of) the pituitary gland

pituitary gland *n* a small endocrine organ attached to the brain that consists of a front lobe and a rear lobe that secrete many important hormones controlling growth, metabolism, etc

placebo *n*, *pl* **placebos** **1** a medication that has no physiological effect and is prescribed more for the mental relief of the patient **2** an inert substance against which an active substance (eg a drug) is tested in a controlled trial

placebo effect *n* improvement in the condition of a sick person that occurs in response to treatment but is prob more connected with mental factors than with the specific treatment

placenta *n*, *pl* **placentas, placentae** **1** a vascular organ in mammals that unites the foetus to the maternal uterus and provides for the nourishment of the foetus and the elimination of waste ☞ REPRODUCTION **2** the part of a flowering plant to which the ovules are attached ☞ FLOWER

placentation *n* the particular type of form and structure of a mammalian or plant placenta

placoid *adj* of or being a scale with an enamel-tipped spine characteristic of cartilaginous fishes

planarian *n* any of a family or order of small cilia-bearing and mostly aquatic flatworms

plankton *n* the floating or weakly

swimming minute animal and plant organisms of a body of water

plant[1] *vt* **1a** to put in the ground, soil, etc for growth <~ *seeds*> **b** to set or sow (land) with seeds or plants **2a** to place (animals) in a new locality **b** to stock with animals ~ *vi* to plant sthg

plant[2] *n* any of a kingdom of living things (eg a green alga, moss, fern, conifer, or flowering plant) typically lacking locomotive movement or obvious nervous or sensory organs ⊚

plantation *n* **1** (a place with) a usu large group of plants, esp trees, under cultivation **2** an agricultural estate, usu worked by resident labour

plant kingdom *n* the one of the 3 basic groups of natural objects that includes all living and extinct plants — compare ANIMAL KINGDOM, MINERAL KINGDOM

plasma *n* **1** the fluid part of blood, lymph, or milk as distinguished from suspended material **2** protoplasm

plasmalemma *n* PLASMA MEMBRANE 1

plasma membrane *n* **1** the semipermeable surface bounding a cell **2** the tonoplast

plasmid *n* a piece of DNA or RNA in some cells, esp bacteria, that exists and reproduces independently of the cell's chromosomes

plasmin *n* an enzyme that breaks down the fibrin of blood clots

plasminogen *n* the substance found in blood plasma and serum from which plasmin is formed

plasmodesma *also* **plasmodesm** *n*, *pl* **plasmodesmata, plasmodesmas** any of the strands of cytoplasm that provide living bridges between some plant cells

plasmodium *n*, *pl* **plasmodia** **1** (an organism consisting of) a (mobile) mass of living matter containing many nuclei and resulting from fusion of amoeba-like cells **2** an individual malaria parasite

plasmolysis *n* shrinking of the cytoplasm away from the wall of a living (plant) cell due to water loss

plastid *n* any of various organelles of plant cells that function as centres of photosynthesis, store starch, oil, etc, or contain pigment

plate *n* an (external) scale or rigid layer of bone, horn, etc forming part of an animal body

platelet *n* BLOOD PLATELET

platyhelminth *n* any of a phylum of soft-bodied flattened worms (eg the planarians, flukes, and tapeworms)

Pleistocene *adj or n* (of or being) the earlier epoch of the Quaternary

pleomorphism *n* the having, assumption, or occurrence of more than 1 distinct form

pleura *n*, *pl* **pleurae, pleuras** the delicate membrane that lines each

ple

half of the thorax of mammals and surrounds the lung of the same side

pleurisy *n* inflammation of the pleura, usu with fever, painful breathing, and oozing of liquid into the pleural cavity

pleuston *n* floating living organisms forming a layer on or near the surface of a body of fresh water

plexus *n* a network of interlacing blood vessels or nerves

plicate *also* **plicated** *adj* folded lengthways like a fan; pleated, ridged <a ~ *leaf*>

Pliocene *adj or n* (of or being) the latest epoch of the Tertiary

ploidy *n* degree of repetition of the haploid number of chromosomes

plumage *n* the entire covering of feathers of a bird

plumose *adj* 1 having feathers or plumes 2 feathery 3 having a main shaft bearing small filaments <*the ~ antennae of an insect*>

plumule *n* the primary bud of a plant embryo

pneumatic *adj* a moved or worked by air pressure b adapted for holding or inflated with compressed air c having air-filled cavities

pneumatophore *n* a muscular gas-containing sac that serves as a float on a hydrozoan colony

pneumococcus *n, pl* **pneumococci** a bacterium that causes acute pneumonia

pneumonia *n* localized or widespread inflammation of the lungs with change from an air-filled to a solid consistency, caused by infection or irritants

pneumonic *adj* 1 of the lungs 2 of or affected with pneumonia

pneumothorax *n* the presence of gas, esp air, in the pleural cavity occurring esp as a result of disease or injury

pod *n* 1 a long seed vessel or fruit, esp of the pea, bean, or other leguminous plant 2 an egg case of a locust or similar insect

podsol *n* podzol

podzol *n* any of a group of soils that have a grey upper layer from which humus and iron and aluminium compounds have leached to enrich the layer below

poikilotherm *n* a living organism (eg a frog) with a variable body temperature usu slightly higher than the temperature of its environment; a cold-blooded organism

point *n* 1 the tip of a projecting body part 2 *pl* (the markings of) the extremities of an animal, esp when of a different colour from the rest of the body

poison¹ *n* 1 a substance that through its chemical action kills, injures, or impairs an organism 2 a substance that inhibits the activity of another substance or the course of a reaction or process <*a catalyst ~*>

poison² *vt* 1 to injure or kill with poison 2 to treat, taint, or impreg-

172

plant 👁

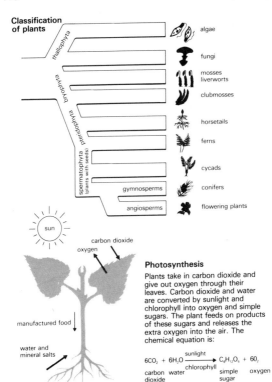

Classification of plants

thallophyta — algae, fungi

bryophyta — mosses, liverworts

pteridophyta — clubmosses, horsetails, ferns

spermatophyta (plants with seeds) — cycads, gymnosperms — conifers, angiosperms — flowering plants

sun

oxygen

carbon dioxide

manufactured food

water and mineral salts

Photosynthesis

Plants take in carbon dioxide and give out oxygen through their leaves. Carbon dioxide and water are converted by sunlight and chlorophyll into oxygen and simple sugars. The plant feeds on products of these sugars and releases the extra oxygen into the air. The chemical equation is:

$$6CO_2 + 6H_2O \xrightarrow[\text{chlorophyll}]{\text{sunlight}} C_6H_{12}O_6 + 6O_2$$

carbon dioxide · water · simple sugar · oxygen

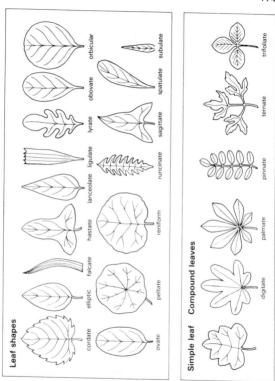

Leaf shapes

orbicular

subulate

obovate

spatulate

lyrate

sagittate

ligulate

runcinate

lanceolate

hastate

reniform

falcate

elliptic

peltate

cordate

ovate

Compound leaves

trifoliate

ternate

pinnate

palmate

digitate

Simple leaf

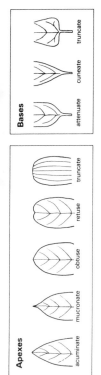

Apexes

acuminate · mucronate · obtuse · retuse · truncate

Bases

attenuate · cuneate · truncate

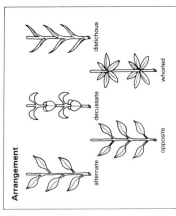

Arrangement

alternate · opposite · decussate · whorled · distichous

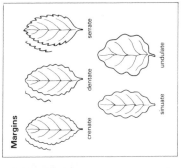

Margins

crenate · dentate · serrate · sinuate · undulate

pol

nate with poison **3** to inhibit the activity, course, or occurrence of

polio *n* poliomyelitis

poliomyelitis *n* an infectious virus disease, esp of children, characterized by inflammation of the nerve cells of the spinal cord, paralysis of the motor nerves, and atrophy of skeletal muscles often with permanent disability and deformity

poll[1] *n* (the hairy top or back of) the head

poll[2] *vt* to cut off or cut short

pollard[1] *n* a tree cut back to the main stem to promote the growth of a dense head of foliage

pollard[2] *vt* to make a pollard of (a tree)

polled *adj* hornless

pollen *n* (a fine dust of) the minute granular spores discharged from the anther of the flower of a flowering plant that serve to fertilize the ovules

pollen basket *n* a smooth area on each hind leg of a bee that serves to collect and transport pollen

pollen tube *n* a tube formed by a pollen grain in contact with the stigma of a flowering plant that conveys the sperm to the ovary

pollinate *vt* to place pollen on the stigma of and so fertilize

polliniferous *adj* producing or (adapted for) bearing pollen

pollinosis, pollenosis *n* hay fever caused by allergic sensitivity to specific pollens

pollute *vt* to make physically impure or unclean; *esp* to contaminate (an environment), esp with man-made waste

pollution *n* **1** polluting or being polluted **2** material that pollutes

polyandrous *adj* having many usu free stamens

polyandry *n* **1** having more than 1 husband at a time — compare POLYGYNY **2** the state of being polyandrous

polychaete *n or adj* (any) of a class of chiefly sea-living annelid worms with many bristles, usu arranged in pairs, along the body — compare OLIGOCHAETE

polygamous, polygamic *adj* **1** having more than 1 mate at a time <*baboons are* ~> **2** bearing both hermaphrodite and unisexual flowers on the same plant

polygenesis *n* origin from more than 1 ancestral line or stock

polygynous *adj* **1** of or practising polygyny **2** *of a plant* having many ovaries

polygyny *n* having more than 1 wife at a time — compare POLYANDRY

polymorphic, polymorphous *adj* having, assuming, or occurring in various forms, characters, or styles

polymorphonuclear leucocyte *n* a granulocyte

polyp *n* **1** a coelenterate with a hollow cylindrical body attached at one end and having a central mouth surrounded by tentacles at the other **2** a projecting mass of tissue (eg a tumour)

polypeptide *n* a long chain of amino acids joined by peptide bonds

polyphagia *n* pathologically excessive appetite or eating

polyphagous *adj* feeding on many kinds of food

polyphyletic *adj* derived from more than 1 ancestral line or more than 1 stock

polyploid *adj* having or being a chromosome number that is a multiple greater than 2 of the haploid number — compare HAPLOID, DIPLOID

polysaccharide *n* a carbohydrate (eg cellulose or starch) consisting of chains of monosaccharide molecules

polyunsaturated *adj*, of a fat or oil rich in unsaturated chemical bonds

polyzoan *n* a bryozoan

pome *n* a fruit (eg an apple) with an outer thickened fleshy layer and a central core with the seeds enclosed in a capsule

pond *n* a body of (fresh) water usu smaller than a lake

pons *n, pl* **pontes** a broad mass of nerve fibres on the lower front surface of the brain

pons Varolii *n* the pons

pontine *adj* of the pons

pooter *n* a portable apparatus for collecting small insects, mites, etc by aspiration

population *n* **1** the total number of individuals inhabiting an area or region **2** a particular group of organisms inhabiting a particular area **3** a set (eg of individual people or items) from which samples are taken for statistical measurement

population explosion *n* a vast usu rapid increase in the size of a living population

pore *n* a minute opening; *esp* one (eg in a membrane, esp the skin, or between soil particles) through which fluids pass or are absorbed

poriferan *n* SPONGE 2

porous *adj* having or full of pores or spaces

portal¹ *n* the point at which sthg (eg a disease-causing agent) enters the body

portal² *adj* **1** of the transverse fissure on the underside of the liver where most of the vessels enter **2** of or being a portal vein

portal vein *n* a vein that transfers blood from one part of the body to another without passing through the heart; *esp* the vein carrying blood from the digestive organs and spleen to the liver

positive *adj* **1** directed or moving towards a source of stimulation <*a ~ response to light*> **2** showing the presence of sthg sought or suspected to be present <*a ~ test for blood*>

posterior *adj* **1** later in time; subsequent **2** situated behind or towards the back: eg **a** *of an animal part near the tail;* caudal **b** *of the human body or its parts* dorsal **3** *of a plant part* (on the side) facing towards the stem or axis; *also* SUPERIOR 1

posthypnotic suggestion *n* the giving of instructions or suggestions to a hypnotized person which he/she will act on when no longer in a trance

postmortem *also* **postmortem examination** *n* an examination of a body after death for determining the cause of death or the character and extent of changes produced by disease

postnatal *adj* subsequent to birth; *also* of or relating to a newborn child

postnuptial *adj* made or occurring after marriage or mating

postpartum *adj* following birth <~ *period*>

postsynaptic *adj* situated or occurring just after a nerve synapse

posture *n* the position or bearing of (relative parts of) the body

potent *adj* **1** chemically or medicinally effective <*a ~ vaccine*> **2** *esp of a male* able to have sexual intercourse

pouch *n* an anatomical structure resembling a pouch: eg **a** a pocket of skin in the abdomen of marsupials for carrying their young **b** a pocket of skin in the cheeks of some rodents used for storing food

poultice[1] *n* a soft usu heated and sometimes medicated mass spread on cloth and applied to inflamed or injured parts (eg sores)

pox *n*, *pl* **pox**, **poxes** **1** a virus disease (eg chicken pox) characterized by eruptive spots **2** syphilis — *infml* **3** *archaic* smallpox

prairie *n* an extensive area of level or rolling (practically) treeless grassland, esp in N America

prebiological *adj* of or being chemical or environmental precursors of the origin of life <~ *molecules*>

Precambrian *adj or n* (of or being) the earliest era of geological history equivalent to the Archaeozoic and Proterozoic eras

precocial *adj*, *of a bird* (having young) capable of a high degree of independent activity from birth <*ducklings are* ~> — compare ALTRICIAL

precocious *adj* **1** exceptionally early in development or occurrence **2** exhibiting mature qualities at an unusually early age

predacious, predaceous *adj* living by preying on other animals; predatory

predation *n* **1** the act of preying or plundering; depredation **2** a mode of life of certain animals in which food is primarily obtained by the killing and consuming of other animals

predatory *adj* living by predation; predacious; *also* adapted to predation

predigest *vt* to prepare (eg food) in an easier form (for consumption)

preen *vt* to trim or dress (as if) with a beak

pregnant *adj* containing unborn young within the body

prehensile *adj* adapted for seizing or grasping, esp by wrapping round <a ~ *tail*>

premature *adj* happening, arriving, existing, or performed before the proper or usual time; *esp, of a human* born after a gestation period of less than 37 weeks

premenstrual *adj* of or occurring in the period just before menstruation <~ *tension*>

premolar *n* the mammalian teeth situated in front of the molars and behind the canines

prenatal *adj* occurring or being in a stage before birth

preoperative *adj* occurring in the period preceding a surgical operation

prepuce *n* the foreskin; *also* a similar fold surrounding the clitoris

preservationist *n* a conservationist

preserve[1] *vt* 1 to keep alive, intact, or free from decay 2 to keep or save from decomposition

preserve[2] *n* an area restricted for the preservation of natural resources (eg animals or trees); *esp* one used for regulated hunting or fishing

pressure point *n* a point where a blood vessel may be compressed against a bone (eg to check bleeding)

presynaptic *adj* situated or occurring just before a nerve synapse

prey[1] *n* 1 an animal taken by a predator as food 2 the act or habit of preying

prey[2] *vi* to seize and devour prey — often + *on* or *upon* <*kestrels* ~ *upon mice*>

primary[1] *adj* 1 of or being formations of the Palaeozoic and earlier periods 2 preparatory to sthg else in a continuing process; elementary <~ *instruction*> 3 belonging to the first group or order in successive divisions, combinations, or ramifications <~ *nerves*> 4 of or being the amino acid sequence in proteins <~ *protein structure*> 5 of, involving, or derived directly from plant-forming tissue, specif meristem, at a growing point <~ *tissue*> <~ *growth*>

primary[2] *n* any of the usu 9 or 10 strong feathers on the joint of a bird's wing furthest from the body ⊃ BIRD

primate *n* any of an order of mammals including human beings, the apes, monkeys, and related forms (eg lemurs and tarsiers) ⊙

primitive *adj* belonging to or characteristic of an early stage of development or evolution <~ *technology*>

primordium *n, pl* **primordia** the rudiment or commencement of a part or organ

probability *n* a measure of the likelihood that a given event will occur, usu expressed as the ratio of the number of times it occurs in a test series to the total number of trials in the series

probability function *n* a function of a discrete random variable that gives the probability that a specified value will occur

proboscidean, proboscidian *n* any of an order of large mammals comprising the elephants and extinct related forms ➤ MAMMAL

proboscis *n*, *pl* **proboscises** *also* **proboscides** 1 a long flexible snout (eg the trunk of an elephant) 2 any of various elongated or extendable tubular parts (eg the sucking organ of a mosquito) of an invertebrate ➤ INSECT

procambium *n* the part of a plant meristem that forms cambium and primary vascular tissues

procaryote *n* a prokaryote

process *n* 1 a natural phenomenon marked by gradual changes that lead towards a particular result <the ~ of growth> 2 a prominent or projecting part of a living organism or an anatomical structure <a bone ~>

procreate *vb* to beget or bring forth (young)

procumbent *adj* being or having

stems that trail along the ground without rooting

producer *n* 1 an individual or entity that grows agricultural products or manufactures articles 2 an organism, usu a photosynthetic green plant, that can synthesize organic matter from inorganic materials and that often serves as food for other organisms – compare CONSUMER ➤ ECOLOGY

proenzyme *n* a zymogen

profile *n* 1 a vertical section of a soil from the ground surface to the underlying material 2 a concise written or spoken biographical sketch

progenitor *n* a biologically ancestral form

progeny *n* offspring of animals or plants

progestational *adj* preceding pregnancy or gestation; *esp* concerning ovulation and corpus luteum formation in female mammals

progesterone *n* a steroid progestational hormone

proglottid *n* a segment of a tapeworm containing both male and female reproductive organs

proglottis *n*, *pl* **proglottides** a proglottid

prognathous *adj* having the jaws projecting beyond the upper part of the face

prognosis *n*, *pl* **prognoses** the prospect of recovery as anticipated

Primate family tree

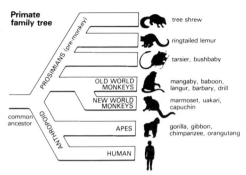

PROSIMIANS (pre-monkey)

tree shrew

ringtailed lemur

tarsier, bushbaby

OLD WORLD MONKEYS — mangaby, baboon, langur, barbary, drill

NEW WORLD MONKEYS — marmoset, uakari, capuchin

ANTHROPOID

common ancestor

APES — gorilla, gibbon, chimpanzee, orangutang

HUMAN

Shape of hand and foot

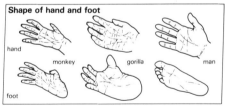

hand

monkey   gorilla   man

foot

Size of brain

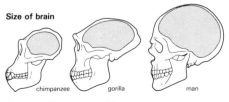

chimpanzee   gorilla   man

pro

from the usual course of disease or peculiarities of a particular case

progressive *adj* increasing in extent or severity <*a ~ disease*>

projection *n* **1** the act of perceiving a subjective mental image as objective **2** the attribution of one's own ideas, feelings, or attitudes to other people or to objects, esp as a defence against feelings of guilt or inadequacy **3** an estimate of future possibilities based on a current trend

prokaryote, procaryote *n* an organism (eg a bacterium or a blue-green alga) that does not have a distinct nucleus — compare EUKARYOTE

prolactin *n* a pituitary hormone that causes milk production in some mammals

prolapse *n* the falling down or slipping of a body part (eg the uterus) from its usual position or relations

proleg *n* a fleshy leg on an abdominal segment of some insect larvae ➔ INSECT

proliferate *vi* to grow or increase (as if) by rapid production of new parts, cells, buds, etc

prolific *adj* producing young or fruit (freely)

promiscuous *adj* not restricted to 1 class or person; indiscriminate; *esp* not restricted to 1 sexual partner

pro-oestrus *n* a period immediately preceding oestrus characterized by preparatory physiological changes

propagate *vt* **1** to reproduce or increase by sexual or asexual reproduction **2** to pass down (eg a characteristic) to offspring ~ *vi* to multiply sexually or asexually

propagation *n* **1** an increase (eg of a type of organism) in numbers **2** the spreading of sthg (eg a belief) abroad or into new regions

prophase *n* **1** the initial phase of mitosis in which chromosomes are condensed from the resting form and split into paired chromatids **2** the initial stage of meiosis in which the chromosomes become visible as paired chromatids and the nuclear membrane disappears

prophylactic *adj* guarding or protecting from or preventing disease

prophylaxis *n, pl* **prophylaxes** measures designed to preserve health and prevent the spread of disease

prostaglandin *n* any of various cyclic fatty acids that are important locally acting hormones in humans and animals and of which one is widely used to induce abortions

prostate, prostate gland *n* a partly muscular, partly glandular body situated around the base of the male mammalian urethra that secretes a major constituent of the ejaculatory fluid ➔ REPRODUCTION

prosthesis *n, pl* **prostheses** an artificial device to replace a missing part of the body

prostomium *n, pl* **prostomia** the portion of the head of various worms and molluscs that is situated in front of the mouth

prostrate *adj* **1** physically exhausted **2** *of a plant* trailing on the ground

protein *n* any of numerous genetically specified naturally occurring extremely complex combinations of amino acids linked by peptide bonds that are essential constituents of all living cells and are an essential part of the diet of animals and humans

proteinase *n* an enzyme that breaks down proteins, esp into peptides

proteolysis *n* the breakdown of proteins or peptides resulting in the formation of simpler (soluble) products

Proterozoic *adj or n* (of or being) an era of geological history between the Archaeozoic and the Palaeozoic

prothallium *n, pl* **prothallia** (a tiny structure of a flowering plant corresponding to) the gamete-producing form of a fern or related plant

prothallus *n* the prothallium

prothrombin *n* a plasma protein produced in the liver in the presence of vitamin K and converted into thrombin in the clotting of blood

protist *n* any of a major group of usu single-celled organisms including bacteria, protozoans, and various algae and fungi

protochordate *n* any of a major division of chordate animals that do not have a vertebral column and include the hemichordates, lancelets, and tunicates

protoplasm *n* **1** the organized complex of organic and inorganic substances (eg proteins and salts in solution) that constitutes the living nucleus, cytoplasm, plastids, and mitochondria of the cell **2** cytoplasm

protoplast *n* the nucleus, cytoplasm, and plasma membrane of a cell as distinguished from nonliving walls and inclusions (eg vacuoles)

prototrophic *adj* deriving nutriment from inorganic sources

protoxylem *n* the first formed xylem developing from the procambium

protozoan *n* any of a phylum or subkingdom of minute single-celled animals which have varied structure and physiology and often complex life cycles

protozoology *n* a branch of zoology dealing with protozoans

protozoon *n, pl* **protozoa** a protozoan

protract *vt* to extend forwards or outwards

protractile *adj* capable of being thrust out <~ *jaws*>

protractor *n* a muscle that extends a body part — compare RETRACTOR

protuberant *adj* thrusting or projecting out from a surrounding or adjacent surface

proximal *adj, esp of an anatomical part* next to or nearest the point of attachment or origin — compare DISTAL

pseudopodium *n, pl* **pseudopodia** a temporary protrusion of a cell (eg an amoeba) that serves to take in food, move the cell, etc

pseudopregnancy *n* 1 FALSE PREGNANCY 2 a state resembling pregnancy that occurs in various mammals usu after an infertile copulation and during which oestrus does not occur

psoriasis *n* a chronic skin condition characterized by distinct red patches covered by white scales

psyche *n* 1 the soul, self 2 the mind

psychedelic *adj* 1 *of drugs* capable of producing altered states of consciousness that involve changed mental and sensory awareness, hallucinations, etc 2 produced by or associated with the use of psychedelic drugs

psychiatry *n* a branch of medicine that deals with mental, emotional, or behavioural disorders

psychic *also* **psychical** *adj* 1 of or originating in the psyche 2 lying outside the sphere of physical science or knowledge 3 sensitive to nonphysical or supernatural forces and influences

psychoactive *adj* affecting the mind or behaviour <~ *drugs*>

psychoanalyse *vt* to treat by means of psychoanalysis

psychoanalysis *n* a method of analysing unconscious mental processes and treating mental disorders, esp by allowing the patient to talk freely about early childhood experiences, dreams, etc

psychodynamics *n* the psychology of mental or emotional forces or processes and their effects on behaviour and mental states; *also* explanation or interpretation (eg of behaviour) in terms of these forces

psycholinguistics *n pl but sing in constr* the study of the interrelation between linguistic behaviour and the minds of speaker and hearer (eg the production and comprehension of speech)

psychological *adj* 1 of psychology 2 mental

psychology *n* 1 the science or study of mind and behaviour 2 the mental or behavioural characteristics of an individual or group

psychometry *n* the psychological theory and technique of the measurement of mental capacities and attributes

psychopath *n* a person suffering from a severe emotional and behavioural disorder characterized by antisocial tendencies and usu the pursuit of immediate gratification through often violent acts;

broadly a dangerously violent mentally ill person

psychopathology *n* (the study of) psychological and behavioural aberrations occurring in mental disorder

psychosexual *adj* of the emotional, mental, or behavioural aspects of sex

psychosis *n, pl* **psychoses** severe mental derangement (eg schizophrenia) that results in the impairment or loss of contact with reality

psychosocial *adj* relating social conditions to mental health <~ *medicine*>

psychosomatic *adj* of or resulting from the interaction of psychological and somatic factors, esp the production of physical symptoms by mental processes <~ *medicine*>

psychotherapy *n* treatment by psychological methods for mental, emotional, or psychosomatic disorders

pteridology *n* the study of ferns

pteridophyte *n* any of a division of plants (eg ferns) that have roots, stems, and leaves but no flowers or seeds ☞ PLANT

ptyalin *n* an enzyme found in the saliva of many animals that breaks down starch into sugar

puberty *n* **1** the condition of being or the period of becoming capable of reproducing sexually **2** the age at which puberty occurs

pubes *n, pl* **pubes** the pubic region or hair

pubescence *n* **1** being pubescent **2** a pubescent covering or surface

pubescent *adj* **1** arriving at or having reached puberty **2** covered with fine soft short hairs — compare HISPID

pubic *adj* of or situated in or near the region of the pubis or the pubic hair

pubic hair *n* the hair that appears at puberty round the genitals

pubis *n, pl* **pubes** the bottom front of the 3 principal bones that form either half of the pelvis

pudendum *n, pl* **pudenda** the external genital organs of a (female) human being — usu pl with sing. meaning

puerperal *adj* of or occurring during (the period immediately following) childbirth

pullulate *vi* to germinate, sprout

pulmonary, pulmonic *adj* of, associated with, or carried on by the lungs

pulmonary artery *n* an artery that conveys deoxygenated blood from the heart to the lungs ☞ RESPIRATION

pulmonary valve *n* the heart valve between the right ventricle and the pulmonary artery that stops blood flowing back into the right ventricle

pulmonary vein *n* a valveless vein that returns oxygenated blood

pul

from the lungs to the heart ⧖ RESPIRATION

pulmonate *adj* **1** having (organs resembling) lungs **2** of a large order of gastropod molluscs having a lung or respiratory sac that includes most land snails and slugs and many freshwater snails

pulp *n* **1** the soft juicy or fleshy part of a fruit or vegetable **2** a soft mass of vegetable matter from which most of the water has been pressed **3** the soft sensitive tissue that fills the central cavity of a tooth ⧖ ANATOMY

pulsate *vi* to beat with a pulse

pulsation *n* rhythmic throbbing or vibrating (eg of an artery); *also* a single beat or throb

pulse[1] *n* the edible seeds of any of various leguminous crops (eg peas, beans, or lentils); *also* the plant yielding these

pulse[2] *n* **1** a regular throbbing caused in the arteries by the contractions of the heart; *also* a single movement of such throbbing **2** the number of beats of a pulse in a specific period of time

pulsimeter *n* an instrument for measuring esp the force and rate of the pulse

punctate *adj* marked with minute spots or depressions <*a ~ leaf*>

pungent *adj* **1** having a stiff and sharp point <*~ leaves*> **2** causing a sharp sensation

pupa *n, pl* **pupae, pupas** the interme-diate usu inactive form of an insect that undergoes metamorphism (eg a bee, moth, or beetle) that occurs between the larva and the imago stages ⧖ LIFE CYCLE

pupate *vi* to become a pupa

pupil *n* the contractile usu round dark opening in the iris of the eye ⧖ SENSE ORGAN

purebred *adj* bred over many generations from members of a recognized breed, strain, or kind without mixture of other blood

purge *vt* to cause evacuation from (eg the bowels)

purine *n* (either of the bases adenine or guanine that are constituents of DNA and RNA and are derivatives of) a compound from which uric acid and related compounds are made in the body

purpura *n* any of several states characterized by patches of purplish discoloration on the skin and mucous membranes and caused by abnormalities in the blood

purulent *adj* **1** containing, consisting of, or being pus <*a ~ discharge*> **2** accompanied by suppuration

pus *n* thick opaque usu yellowish white fluid matter formed by suppuration (eg in an abscess)

pustular *adj* of, resembling, or covered with pustules

pustule *n* **1** a small raised spot on the skin having an inflamed base and containing pus **2** a small raised area like a blister or pimple

putrefaction *n* the decomposition of organic matter; *esp* the breakdown of proteins by bacteria and fungi, typically in the absence of oxygen, with the formation of foul-smelling incompletely oxidized products

putrefy *vb* to make or become putrid

putrescent *adj* of or undergoing putrefaction

putrid *adj* **1** in a state of putrefaction **2** (characteristic) of putrefaction; *esp* foul-smelling

pylorus *n, pl* **pylori** the opening from the vertebrate stomach into the intestine

pyrenoid *n* any of the protein bodies that act as centres for starch deposition in some algae and other lower organisms

pyretic *adj* of fever

pyrexia *n* abnormal elevation of body temperature

pyridoxine *also* **pyridoxin** *n* one of the vitamin B₆ group found esp in cereal foods and convertible in the body into phosphate compounds that are important coenzymes

pyrimidine *n* any of the bases cytosine, thymine, or uracil that are constituents of DNA and RNA

pyruvic acid *n* a liquid organic acid that smells like vinegar and is an important intermediate compound in metabolism and fermentation

pyxidium *n, pl* **pyxidia** a capsular

fruit that opens at maturity with the upper part falling off like a cap

quadriceps *n* the large muscle at the front of the thigh that acts to straighten the leg at the knee joint

quadriplegic *n* affected with paralysis of both arms and both legs

quadrumana *n pl* primates, excluding human beings, considered as a group distinguished by hand-shaped feet

quadruped *n* an animal having 4 feet

quagmire *n* soft miry land that shakes or yields under the foot

quarantine *n* **1** (the period of) a restraint on the activities or communication of people or the transport of goods or animals, designed to prevent the spread of disease or pests **2** a place in which people, animals, vehicles, etc under quarantine are kept

quartile *n* any of 3 numbers that divide a frequency distribution into 4 equal intervals

Quaternary *adj* of or being the geological period from the end of the Tertiary to the present time

queen *n* the fertile fully developed female in a colony of bees, ants, or termites

queen substance *n* a pheromone secreted by queen bees that is consumed by worker bees and inhibits the development of their ovaries

quill *n* **1** a roll of dried bark <*cinna-*

mon ~s> **2** the hollow horny barrel of a feather **3** any of the large stiff feathers of a bird's wing or tail **4** any of the hollow sharp spines of a porcupine, hedgehog, etc

quinine *n* a substance with a bitter taste that is obtained from cinchona bark, is used as a tonic, and was formerly the major drug in the treatment of malaria

quotient *n* the ratio, usu multiplied by 100, between a test score and a measurement on which that score might be expected largely to depend — compare INTELLIGENCE QUOTIENT

rabies *n, pl* **rabies** a fatal short-lasting virus disease of the nervous system of warm-blooded animals, transmitted esp through the bite of an affected animal, and characterized by extreme fear of water and convulsions

race *n* **1** an actually or potentially interbreeding group within a species; *also* a category (eg a subspecies) in classification representing such a group **2** a division of mankind having traits that are transmissible by descent and sufficient to characterize it as a distinct human type

raceme *n* a simple stalk of flowers (eg that of the lily of the valley) in which the flowers are borne on short side-stalks of about equal length along an elongated main stem ➮ FLOWER

racemose *adj* having or growing in the form of a raceme

rachis *n, pl* **rachises** *also* **rachides 1** SPINAL COLUMN **2a(1)** the main stem of a plant's inflorescence **(2)** an extension of the stalk of a compound leaf that bears the leaflets **b** the part of the shaft of a feather that bears the barbs

rachitis *n* rickets

radial[1] *adj* **1** (having parts) arranged like rays or radii from a central point or axis **2** of or situated near a radius bone (eg in the human forearm)

radial[2] *n* a radial body part (eg an artery)

radial symmetry *n* the condition of having similar parts symmetrically arranged around a central axis

radiate *adj* having rays or radial parts; *specif* having radial symmetry

radiation sickness *n* sickness that results from overexposure to ionizing radiation (eg Xrays), commonly marked by fatigue, nausea, vomiting, loss of teeth and hair, and, in more severe cases, leukaemia

radical *adj* **1** of or growing from the root or the base of a stem — compare CAULINE **2** designed to remove the root of a disease or all diseased tissue <~ *surgery*>

radices *pl of* radix

radicle *n* **1** the lower part of the axis of a plant embryo or seedling, including the embryonic root **2** the rootlike beginning of an anatomical vessel or part

radioactive *adj* emitting ionizing radiation by the disintegration of the nuclei of atoms

radiocarbon *n* radioactive carbon; *esp* CARBON 14

radiogram *n* a radiograph

radiograph *n* a picture produced on a sensitive surface by a form of radiation other than light; *specif* an X-ray or gamma-ray photograph

radioisotope *n* a radioactive isotope

radiolarian *n* any of a large order of marine protozoans with a skeleton made of silica and radiating threadlike pseudopodia

radiology *n* the study and use of radioactive substances and high-energy radiations; *esp* the use of radiant energy (eg X rays and gamma rays) in the diagnosis and treatment of disease

radiotherapy *n* the treatment of disease (eg cancer) by means of X rays or radiation from radioactive substances

radio-ulna *n* a bone in the forelimb of an amphibian (eg a frog) that represents the fused radius and ulna of less primitive vertebrate animals (eg mammals)

radius *n, pl* **radii** *also* **radiuses** the bone on the thumb side of the human forearm; *also* a corresponding part in forms of vertebrate

animals higher than fishes ☞ ANATOMY

radix *n, pl* **radices, radixes** a root or rootlike part

radula *n, pl* **radulae** *also* **radulas** a horny band covered with minute teeth found in some molluscs (eg snails) and used to tear up food and draw it into the mouth

rain forest *n* a dense tropical woodland with an annual rainfall of at least 2500mm (about 100in) and containing lofty broad-leaved evergreen trees forming a continuous canopy

ramification *n* **1a** the act or process of branching out **b** the arrangement of branches (eg on a plant) **2** a branched structure

ramiform *adj* resembling or constituting branches

ramify *vb* to (cause to) separate or split up into branches, divisions, or constituent parts

ramose *adj* consisting of or having branches

random *adj* (of, consisting of, or being events, parts, etc) having or relating to a probability of occurring equal to that of all similar parts, events, etc

range¹ *n* **1** the region throughout which a kind of living organism or ecological community naturally lives or occurs **2** (the difference between) the least and greatest values of an attribute or series

range² *vi* to live, occur in, or be native to, a specified region

rank¹ *adj* (covered with vegetation which is) excessively vigorous and often coarse in growth

rank² *n* a row or series

rank correlation *n* a measure of correlation based on rank

rape *n* a European plant of the mustard family grown as a forage crop and for its seeds which yield rapeseed oil

rapid eye movement *n* rapid movement of the eyes that occurs during the phases of sleep when dreaming is taking place

raptorial *adj* 1 *esp of a bird* predatory 2 *of birds' feet* adapted for seizing prey 3 of or being a bird of prey

rash *n* an outbreak of spots on the body

ratoon¹ *n* a new shoot that develops from the root of the sugarcane or other perennial plant after cropping

ratoon² *vi* to sprout from the root

ray¹ *n* any of numerous cartilaginous fishes having the eyes on the upper surface of a flattened body and a long narrow tail

ray² *n* 1 any of the bony rods that support the fin of a fish 2 any of the radiating parts of the body of a radially symmetrical animal (eg a starfish) 3 RAY FLOWER 4 MEDULLARY RAY

ray flower *n* any of the strap-shaped florets forming the outer ring of the head of a composite plant (eg an aster or daisy) having central disc florets

react *vi* to respond to a stimulus

reaction *n* 1 the response of tissues to a foreign substance (eg an antigen or infective agent) 2 a mental or emotional response to circumstances

reactive *adj* tending to or liable to react <*highly ~ chemicals*>

reason *n* 1a the power of comprehending, inferring, or thinking, esp in orderly rational ways; intelligence b proper exercise of the mind 2 sanity <*lost his ~*>

recapitulation *n* the supposed occurrence in the development of an embryo of successive stages resembling the series of ancestral types from which the organism has evolved

Recent *adj* of or being the present or post-Pleistocene geological epoch

receptacle *n* the end of the flower stalk of a flowering plant upon which the floral organs are borne ☞ FLOWER

receptive *adj* able to receive and transmit stimuli; sensory

receptor *n* 1 a cell or group of cells that receives stimuli; SENSE ORGAN 2 a molecule or group of molecules, esp on the surface of a cell, that have an affinity for a particular chemical (eg a neurotransmitter)

recessive *adj* being the one of a pair of (genes determining) contrasting inherited characteristics that is suppressed if a dominant gene is present — compare ¹DOMINANT 2

reclaim *vt* to make available for human use by changing natural conditions <~ed *marshland*>

recombinant *adj* 1 exhibiting genetic recombination <~ *progeny*> 2 of or being DNA prepared in the laboratory by combining pieces of DNA from several different species of organisms

recombination *n* the formation of new combinations of genes in progeny that did not occur in the parents

rectal *adj* of, affecting, or near the rectum

rectrix *n, pl* **rectrices** any of a bird's tail feathers that are important in controlling flight direction

rectum *n, pl* **rectums, recta** the last part of the intestine of a vertebrate, ending at the anus

recurrent *adj, esp of nerves and anatomical vessels* running or turning back in a direction opposite to a former course

recycle *vt* to pass through a series of changes or treatments so as to return to a previous stage in a cyclic process; *specif* to process (sewage, waste paper, glass, etc) for

conversion back into a useful product

red alga *n* any of a division of algae that are seaweeds with a predominantly red colour

red blood cell, red cell *n* any of the haemoglobin-containing cells that carry oxygen to the tissues and are responsible for the red colour of vertebrate blood — compare WHITE BLOOD CELL

red blood corpuscle, red corpuscle *n* RED BLOOD CELL

redia *n, pl* **rediae** *also* **redias** a larva of any of various parasitic trematode worms that either produces another generation of rediae or develops into a cercaria

redox *adj* of or involving both oxidation and reduction <*a* ~ *reaction*>

reducing agent *n* a substance that reduces a chemical compound usu by donating electrons

reductant *n* REDUCING AGENT

reduction division *n* (the first division of) meiosis of cells

reflex¹ *n* 1 an automatic response to a stimulus that does not reach the level of consciousness 2 *pl* the power of acting or responding with adequate speed

reflex² *adj* 1 bent, turned, or directed back <*a stem with* ~ *leaves*> 2 directed back upon the mind or its operations; introspective 3 of, being, or produced by a reflex

ref

without intervention of consciousness

reflex arc *n* the complete nervous path involved in a reflex

reflexed *adj* bent or curved backwards or downwards <~ *petals*>

refractory *adj* 1 resistant to treatment or cure <*a ~ cough*> 2 immune <*after recovery they were ~ to infection*>

regenerate *vi, of a body or body part* to undergo renewal or regrowth (eg after injury) ~ *vt* to generate or produce anew; *esp* to replace (a body part) by a new growth of tissue

region *n* 1 an area characterized by a particular flora or fauna 2 an indefinite area surrounding a specified body part <*the abdominal ~*>

regress *vi* to tend to approach or revert to a mean ~ *vt* to induce, esp by hypnosis, a state of psychological regression in

regression *n* 1 reversion to an earlier mental or behavioural level 2 the statistical analysis of the association between 2 or more variables, esp so that prediction can be made

regulator gene *n* a gene that controls the production of a repressor

regular *adj* perfectly (radially) symmetrical or even

regurgitate *vb* to vomit or pour back or out (as if) from a cavity

rehabilitate *vt* to restore to a condition of health or useful and con-

structive activity (eg after illness or imprisonment)

reinforce *vt* to stimulate (an experimental subject) with a reward following a correct or desired performance; *also* to encourage (a response) with a reward

reject *vt* 1 to eject; *esp* ²VOMIT 2 to fail to accept (eg a skin graft or transplanted organ) as part of the organism because of immunological differences

relative *n* an animal or plant related to another by common descent

relaxin *n* a hormone produced by the corpus luteum in the ovary of a pregnant mammal that makes birth easier by causing relaxation of the pelvic ligaments

relict *n* a (type of) plant or animal that is a remnant of an otherwise extinct flora, fauna, or kind of organism

REM *n* RAPID EYE MOVEMENT

remedial *adj* intended as a remedy <~ *treatment*>

remedy *n* a medicine, application, or treatment that relieves or cures a disease

renal *adj* relating to, involving, or located in the region of the kidneys

reniform *adj* kidney-shaped ⟹ PLANT

rennin *n* any of several enzymes that coagulate milk and are used in making cheese and junkets; *esp* one from the mucous membrane of the stomach of a calf

repent *adj, of a plant part* creeping, prostrate

replicate *vt* **1** to duplicate, repeat <~ *a statistical experiment*> **2** to fold or bend back

repress *vt* **1** to exclude (eg a feeling) from consciousness by psychological repression — compare SUP-PRESS 1 **2** to inactive (a gene) by blocking

repression *n* a psychological process by which unacceptable desires or impulses are excluded from conscious awareness

repressor *n* a product of a regulator gene that represses the function of an operator gene

reproduce *vt* to produce (new living things of the same kind) by a sexual or asexual ~ *vi* process to produce offspring

reproduction *n* the act or process of reproducing; *specif* the sexual or asexual process by which plants and animals give rise to offspring ⊚

reptile *n* any of a class of air-breathing vertebrates that include the alligators and crocodiles, lizards, snakes, turtles, and extinct related forms (eg the dinosaurs) and have a bony skeleton and a body usu covered with scales or bony plates ⬦ EVOLUTION

reptilian *adj* **1** resembling or having the characteristics of a reptile **2** of the reptiles

resident *adj, of an animal* not migratory

residual *n* the difference between **a** results obtained by observation and by computation from a formula **b** the mean of several observations and any one of them

resorb *vt* to swallow, suck in, or absorb again ~ *vi* to undergo resorption

resorption *n* resorbing, esp of distinct tissues in the body, or being resorbed

respiration *n* **1a** the process by which air or dissolved gases are brought into intimate contact with the circulating medium of a multi-cellular organism (eg by breathing) ⊚ **b** (a single complete act of) breathing **2** the processes by which an organism supplies its cells with the oxygen needed for metabolism and removes the carbon dioxide formed in energy-producing reactions **3** any of various energy-yielding reactions involving oxidation that occur in living cells

respirator *n* **1** a device worn over the mouth or nose to prevent the breathing of poisonous gases, harmful dusts, etc **2** a device for maintaining artificial respiration

respiratory pigment *n* any of various proteins that function in the transfer of oxygen in cellular respiration

respire *vi* **1** to breathe **2** *of a cell or tissue* to take up oxygen and pro-

Woman

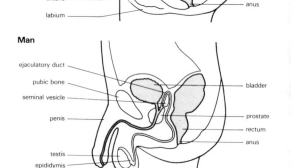

ovary
peritoneal cavity
bladder
pubic bone
vagina
clitoris
labium

fallopian tube
womb (uterus)
neck of womb (cervix)
rectum
anus

Man

ejaculatory duct
pubic bone
seminal vesicle
penis
testis
epididymis

bladder
prostate
rectum
anus
scrotum

In a woman, one ovary produces an egg (ovum) approximately every 28 days. In a man, spermatozoa are produced in the testes and travel up the epididymis to the seminal vesicles where they are stored in fluid produced by the prostate.

Fertilization

When an egg is released from the ovary, it travels down the fallopian tube and may be fertilized by a spermatozoon during copulation. If fertilization takes place the egg usually implants in the lining of the womb. An egg takes several days to travel from the ovary to the womb.

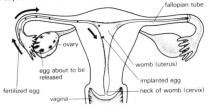

fallopian tube

ovary

egg about to be released

womb (uterus)

implanted egg

fertilized egg

neck of womb (cervix)

vagina

A nine-month-old foetus

When the fertilized egg starts to grow and develop, it is known as an embryo. After 3 months the embryo looks like a human and it is then called a foetus. The embryo/foetus absorbs food and oxygen from its mother through the placenta and umbilical cord.

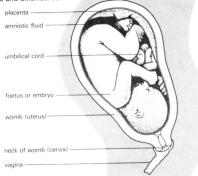

placenta

amniotic fluid

umbilical cord

foetus or embryo

womb (uterus)

neck of womb (cervix)

vagina

The respiratory system

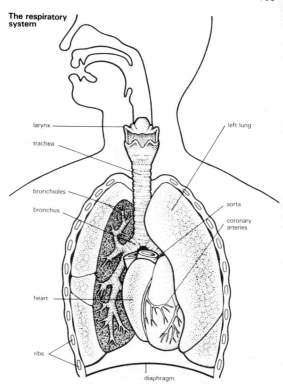

larynx

trachea

bronchioles

bronchus

heart

ribs

left lung

aorta

coronary arteries

diaphragm

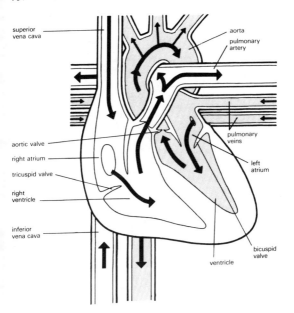

The heart

The right heart (right atrium and right ventricle) pumps deoxygenated blood
through the pulmonary artery to the lungs. The left heart (left atrium and
left venticle) pumps oxygenated blood through the aorta to the body.
Arteries carry blood (usually oxygenated) away from the heart, veins carry
blood (usually deoxygenated) to the heart.

duce carbon dioxide during respiration ~ *vt* to breathe

response *n* a change in the behaviour of an organism resulting from stimulation

retarded *adj* slow in intellectual or emotional development or academic progress

retention *n* abnormal retaining of a fluid (eg urine) in a body cavity

reticulate, reticular *also* **reticulose** *adj* resembling a net; *esp* having veins, fibres, or lines crossing

reticulocyte *n* a young non-nucleated red blood cell

reticulum *n* 1 the second stomach of a ruminant mammal in which folds of the lining form hexagonal cells 2 a reticulate formation; a network

retina *n, pl* **retinas, retinae** the sensory membrane at the back of the eye that receives the image formed by the lens and is connected with the brain by the optic nerve ➔ SENSE ORGAN

retinal *n* a derivative of vitamin A that in combination with proteins forms the visual pigments of the retinal rods and cones

retinol *n* the chief and typical vitamin A

retract *vi* to draw back or in

retractor *n* 1 a surgical instrument for holding open the edges of a wound 2 a muscle that draws in a body part — compare PROTRACTOR

retrogression *n* a reversal in development or condition; *esp* a return to a less advanced or specialized state during the development of an organism

retrorse *adj* bent backwards or downwards

retuse *adj, of a leaf* having a rounded and notched end ➔ PLANT

reversion *n* (an organism showing) a return to an ancestral type or reappearance of an ancestral character

revolute *adj* rolled backwards or downwards <*a leaf with ~ margins*>

rhesus factor *n* any of several antigens in red blood cells that can induce intense allergic reactions

rheumatism *n* 1 any of various conditions characterized by inflammation and pain in muscles, joints, or fibrous tissue 2 RHEUMATOID ARTHRITIS

rheumatoid arthritis *n* painful inflammation and swelling of joint structures occurring as a progressively worsening disease of unknown cause

rhinal *adj* nasal

rhinitis *n* inflammation of the mucous membrane of the nose

rhizocarp *n* a plant with perennial underground parts but annual stems and foliage

rhizome *n* an elongated (thickened and horizontal) underground plant stem distinguished from a true root in having buds and usu scalelike leaves

rhizopod *n* any of a subclass of related protozoans (eg an amoeba) with lobed rootlike pseudopodia

rhodopsin *n* a light-sensitive pigment in the retinal rods of marine fishes and most higher vertebrates whose presence determines the sensitivity of the rods to differing intensities of illumination — compare IODOPSIN

rhythm *n* **1** a regularly recurrent change in a biological process or state (eg with night and day) **2** **rhythm, rhythm method** birth control by abstinence from sexual intercourse during the period when ovulation is most likely to occur

rib *n* **1** any of the paired curved rods of bone or cartilage that stiffen the body walls of most vertebrates and protect the heart, lungs, etc ⊃ ANATOMY **2** a vein of a leaf or insect's wing

rib cage *n* the enclosing wall of the chest consisting chiefly of the ribs and their connections

riboflavin, riboflavine *n* a yellow vitamin of the vitamin B complex occurring esp in milk and liver

ribonucleic acid *n* RNA

ribonucleotide *n* a nucleotide containing ribose rather than deoxyribose and occurring esp as a constituent of RNA

ribose *n* a pentose sugar occurring esp in ribonucleotides

ribosome *n* any of the minute granules containing RNA and protein that occur in cells and are the sites where proteins are synthesized

rickets *n* soft and deformed bones in children caused by failure to assimilate and use calcium and phosphorus, normally due to a lack of sunlight or vitamin D

rickettsia *n, pl* **rickettsias, rickettsiae** any of a family of microorganisms similar to bacteria that are intracellular parasites and cause various diseases (eg typhus)

right atrioventricular valve *n* TRICUSPID VALVE

rigor mortis *n* the temporary rigidity of muscles that occurs after death

ringed *adj* encircled or marked (as if) with rings

ringworm *n* any of several contagious fungous diseases of the skin, hair, or nails in which ring-shaped discoloured blister-covered patches form on the skin

RNA *n* any of various nucleic acids similar to DNA that contain ribose and uracil as structural components instead of deoxyribose and thymine, and are associated with the control of cellular chemical activities

rod *n* any of the relatively long rod-shaped light receptors in the retina

rod

that are sensitive to faint light — compare CONE 2

rodent *n* any of an order of relatively small gnawing mammals including the mice, rats, squirrels, and beavers ☞ MAMMAL

roe *n* **1** the eggs of a female fish, esp when still enclosed in a membrane, or the corresponding part of a male fish **2** the eggs or ovaries of an invertebrate (eg a lobster)

rogue¹ *n* sby or sthg that displays a chance variation making it inferior to others

rogue² *vb* **roguing, rogueing** to weed out inferior, diseased, etc plants (from)

rogue³ *adj,* (roaming alone and) vicious and destructive <*a ~ elephant*>

role, rôle *n* a socially expected behaviour pattern, usu determined by an individual's status in a particular society

roost¹ *n* **1** a support or place where birds roost **2** a group of birds roosting together

roost² *vi, esp of a bird* to settle down for rest or sleep; perch

root¹ *n* **1** the (underground) part of a flowering plant that usu anchors and supports it and absorbs and stores food **2a** the end of a nerve nearest the brain and spinal cord **b** the part of a tooth, hair, the tongue, etc by which it is attached to the body ☞ ANATOMY

root² *vt* **1** to give or enable to develop roots **2** to fix or implant (as if) by roots ~ *vi* to grow roots or take root

root cap *n* a protective cap of cells that covers the growing point at the end of most root tips

root crop *n* a crop (eg turnips or sugar beet) grown for its enlarged roots

root hair *n* an outgrowth of an epidermal cell near the root tip that functions in absorption of water and minerals

root-mean-square *n* the square root of the arithmetic mean of the squares of a set of numbers

rootstock *n* **1** an underground plant part formed from several stems **2** a stock for grafting consisting of (a piece of) root

Rorschach test *n* a personality test based on the interpretation of sby's reactions to a set of standard inkblot designs

rosaceous *adj* of or belonging to the rose family of plants

rosette *n* **1** a rosette-shaped structure or marking on an animal **2** a cluster of leaves in crowded circles or spirals (eg in the dandelion)

rostellum *n* a small beaklike body part; a small rostrum

rostrum *n, pl* **rostrums, rostra** a body part (eg an insect's snout or beak) shaped like a bird's bill

rot¹ *vb* **-tt-** *vi* to undergo decomposition, esp from the action of bacteria or fungi — often + *down* ~ *vt* to cause to decompose or deteriorate

rot² *n* **1** (sthg) rotting or being rotten; decay **2** any of several plant or animal diseases with breakdown and death of tissues

rotate *adj*, *of a flower* with petals or sepals radiating like the spokes of a wheel

rotation *n* **1** the growing of different crops in succession in 1 field, usu in a regular sequence **2** the turning of a limb about its long axis

rotifer *n* any of a class or phylum of minute aquatic invertebrate animals with circles of cilia at the front that look like rapidly revolving wheels

rotten *adj* having rotted; putrid

roughage *n* coarse bulky food (eg bran) that is relatively high in fibre and low in digestible nutrients and that by its bulk stimulates intestinal peristalsis

roundworm *n* a nematode

royal jelly *n* a highly nutritious secretion of the honeybee that is fed to the very young larvae and to all larvae that will develop into queens

rubella *n* GERMAN MEASLES

rudiment *n* **1** a deficiently developed body part or organ; VESTIGE **2 2** a primordium

rudimentary *adj* very poorly developed or represented only by a vestige <*the ~ tail of a hyrax*>

rumen *n*, *pl* **rumina, rumens** the large first compartment of the stomach of a ruminant mammal in which cellulose is broken down, esp by the action of symbiotic bacteria

ruminant *adj* **1** that chews the cud **2** of or being (a member of) a suborder of hoofed mammals including the cattle, sheep, giraffes, and camels that chew the cud and have a complex 3- or 4-chambered stomach

run¹ *vi* **-nn-; ran; run 1** *of fish* to migrate or move in schools; *esp* to ascend a river to spawn **2** to discharge pus or serum <*a ~ning sore*>

run² *n* **1** (a school of fish) migrating or ascending a river to spawn **2** a way, track, etc frequented by animals

runcinate *adj*, *of a leaf* having large downward-pointing teeth <*the ~ leaves of the dandelion*> ⤳ PLANT

runner *n* a stolon

runt *n* an animal unusually small of its kind; *esp* the smallest of a litter

rupture¹ *n* the tearing apart of a tissue, esp muscle a hernia

rupture² *vt* to produce a rupture in ~ *vi* to have or undergo a rupture

rural *adj* of the country, country people or life, or agriculture

rust *n* (a fungus causing) any of numerous destructive diseases of plants esp cereals in which reddish brown pustular lesions form

rye *n* (the seeds, from which a

wholemeal flour is made, of) a hardy grass widely grown for grain

sac *n* a (fluid-filled) pouch within an animal or plant

saccate *adj* having the form of a sac or pouch <*a ~ corolla*>

saccharide *n* SUGAR 2

saccharine *adj* of, like, or containing sugar <*~ taste*>

saccharometer *n* a device for measuring the amount of sugar in a solution

saccharose *n* sucrose

saccular *adj* resembling a sac

sacculate, sacculated *adj* having or formed of a series of saclike expansions

saccule *n* a little sac; *specif* the smaller chamber of the membranous labyrinth of the ear ⮕ SENSE ORGAN

sacculus *n, pl* **sacculi** a saccule

sacral *adj* of or lying near the sacrum

sacrum *n, pl* **sacra** the part of the vertebral column that is directly connected with or forms part of the pelvis and in humans consists of 5 united vertebrae ⮕ ANATOMY

saddle *n* a saddle-shaped marking on the back of an animal

safe period *n* the time during or near the menstrual period when conception is least likely to occur

sagittal *adj* 1 of the join between the parietal bones that stretches from the front to the back of the top of the skull 2 of, situated in, or being

(a plane parallel to) the middle plane or midline of the body

sagittate *adj, of a plant or animal part, esp a leaf* shaped like an arrowhead ⮕ PLANT

salientian *n* any of an order of amphibians including the frogs and toads, that lack a tail as adults and have long hind limbs suited to leaping and swimming

salina *n* a salt marsh, lake, spring, etc

saline[1] *adj* of or containing salt <*a ~ solution*>

saline[2] *n* a saline solution; *esp* one similar in concentration to body fluids

saliva *n* a slightly alkaline mixture of water, protein, salts, and often enzymes that is secreted into the mouth by glands, and that lubricates ingested food and often begins the breakdown of starches

salivate *vi* to have an (excessive) flow of saliva

Salk vaccine *n* a vaccine against polio

salmonella *n, pl* **salmonellae, salmonellas, salmonella** any of a genus of bacteria that cause diseases, esp food poisoning, in warm-blooded animals

salpinx *n, pl* **salpinges** 1 EUSTACHIAN TUBE 2 FALLOPIAN TUBE

salt[1] *n* 1 sodium chloride 2 *pl* a mixture of the salts of alkali metals or magnesium (eg Epsom salts) used as a purgative 3 any of numerous compounds formed by the (partial) replacement of the hydrogen ion of an acid by a (radical acting like a) metal

salt² *adj* being or inducing a taste similar to that of common salt that is one of the 4 basic taste sensations — compare BITTER, SOUR, SWEET ☞ SENSE ORGAN

saltwater *adj* of, living in, or being salt water

salve *n* an ointment for application to wounds or sores

samara *n* a dry indehiscent usu 1-seeded winged fruit (eg of a sycamore)

sample¹ *n* a part of a statistical population whose properties are studied to gain information about the whole

sample² *vt* **sampling** to take a sample of or from; *esp* to test the quality of by a sample

sampling *n* **1** a small (statistical) sample **2** the act, process, or technique of selecting a suitable sample

sanctuary *n* a refuge for (endangered) wildlife where predators are controlled and hunting is illegal <*a bird* ~>

sand *n* loose granular particles smaller than gravel and coarser than silt that result from the disintegration of (silica-rich) rocks

sane *adj* (produced by a mind that is) mentally sound; able to antici-

pate and appraise the effect of one's actions

sanguineous *adj* of or containing blood

sanity *n* being sane; *esp* soundness or health of mind

sap¹ *n* a watery solution that circulates through a plant's vascular system

sap² *vt* **-pp-** to drain or deprive of sap

sapling *n* a young tree

sappy *adj* resembling or consisting largely of sapwood

saprogenic *adj* of, causing, or resulting from putrefaction

saprophagous *adj* feeding on decaying matter

saprophytic *adj*, *esp of a plant* obtaining food by absorbing the products of organic breakdown and decay or other dissolved organic material

sapwood *n* the younger softer usu lighter-coloured living outer part of wood that lies between the bark and the heartwood

sarcoma *n*, *pl* **sarcomas, sarcomata** a cancer arising in connective tissue, bone, or muscle

sarcoplasm *n* the cytoplasm of a striated muscle fibre

sargasso *n*, *pl* **sargassos** a large mass of floating vegetation, esp sargassums, in the sea

sargassum *n* any of a genus of floating seaweeds that have air bladders

sartorius n, pl **sartorii** a long muscle that crosses the front of the thigh obliquely

satyriasis n excessive sexual desire in a male — compare NYMPHOMANIA

saurian n any of a group of reptiles including the lizards and formerly the crocodiles and dinosaurs

savanna, savannah n a tropical or subtropical grassland with scattered trees

saxicolous, saxicoline adj inhabiting or growing among rocks <~ lichens>

scab n 1 scabies of domestic animals 2 a crust of hardened blood and serum over a wound 3 any of various plant diseases characterized by crusted spots; also any of these spots

scabies n, pl **scabies** a skin disease, esp contagious itch or mange, caused by a parasitic mite and usu characterized by oozing scabs

scabrous adj rough to the touch with scales, scabs, raised patches, etc

scalar adj 1 having an uninterrupted series of steps <~ cells> 2 capable of being represented by a point on a scale <a ~ quantity>

scale n 1 (a small thin plate resembling) a small flattened rigid plate forming part of the external body covering of a fish, reptile, etc 2 a small thin dry flake shed from the skin 3 a usu thin, membranous,

chaffy, or woody modified leaf 4 infestation with or disease caused by scale insects

scale insect n any of numerous small insects with scale-like females attached to the host plant and young that suck plant juices

scale leaf n a modified usu small and scaly leaf (eg of a cypress)

scalp n (the part of a lower mammal corresponding to) the skin of the human head, usu covered with hair in both sexes

scaly adj covered with or composed of scale or scales

scan vt **-nn-** to make a detailed examination of (eg the human body) using any of a variety of sensing devices (eg ones using ultrasonics, thermal radiation, X rays, or radiation from radioactive materials)

scape n 1 a leafless flower stalk arising directly from the root of a plant (eg in the dandelion) 2 the shaft of an animal part (eg an antenna or feather)

scaphoid[1] adj navicular

scaphoid[2] n the navicular of the carpus or tarsus

scapula n, pl **scapulae, scapulas** a large flat triangular bone at the upper part of each side of the back forming most of each half of the shoulder girdle: the shoulder blade
☞ ANATOMY

scapular[1] n any of the feathers covering the base of a bird's wing

scapular² *adj* of the shoulder, the shoulder blade, or scapular feathers

scar¹ *n* **1** a mark left (eg on the skin) by the healing of injured tissue **2** CICATRIX 2 **3** a lasting moral or emotional injury

scar² *vb* **-rr-** *vt* **1** to mark with a scar **2** to do lasting injury to ~ *vi* **1** to form a scar **2** to become scarred

scarify *vt* **1** to make scratches or small cuts in (eg the skin) **2** to break up and loosen the surface of (eg a field or road)

scarlet fever *n* an infectious fever caused by a streptococcus in which there is a red rash and inflammation of the nose, throat, and mouth

scatology *n* the biologically oriented study of excrement (eg for the determination of diet)

scatter *n* **1** a small supply or number irregularly distributed **2** the state or extent of being scattered

scavenge *vt* to feed on (carrion or refuse) ~ *vi* **1** to search for reusable material **2** to obtain food by scavenging <*dogs* scavenging *on kitchen waste*>

scavenger *n* an organism that feeds on refuse or carrion ⋺ ECOLOGY

schistosome *n* any of a family of elongated trematode worms that parasitize the blood vessels of birds and mammals

schistosomiasis *n, pl* **schistosomiases** a severe endemic disease of human beings in much of Asia, Africa, and S America marked esp by blood loss and tissue damage

schizocarp *n* a dry compound fruit that splits into several indehiscent single-seeded parts

schizoid *adj* characterized by, resulting from, tending towards, or suggestive of schizophrenia

schizomycete *n* a bacterium

schizophrenia *n* a mental disorder characterized by loss of contact with reality and disintegration of personality, usu with hallucinations and disorder of feeling, behaviour, etc

Schwann cell *n* a cell whose plasma membrane forms the myelin sheath of a nerve fibre

sciatic *adj* **1** of or situated near the hip **2** of or caused by sciatica <~ *pains*>

sciatica *n* pain in the back of the thigh, buttocks, and lower back caused esp by pressure on the sciatic nerve

sciatic nerve *n* either of the 2 largest nerves in the body that pass out of the pelvis and down the back of the thigh, one on each side of the body, and supply the pelvic region and leg

scintigraphy *n* the production of a two-dimensional picture of a body part by detection of the emitted radiation after administration of a radioisotope

scion *n* a detached living part of a plant that is joined to a stock in grafting and usu supplying parts above ground of the resulting graft

scirrhus *n, pl* **scirrhi** a hard slow-growing malignant tumour, esp in the breast, consisting mostly of fibrous tissue

sclera *n* the opaque white outer coat enclosing the eyeball except for the part covered by the cornea ☞ SENSE ORGAN

sclerenchyma *n* a supporting tissue in higher plants composed of cells with thickened and woody walls — compare COLLENCHYMA, PARENCHYMA

sclerosis *n* **1** (a disease characterized by) abnormal hardening of tissue, esp from overgrowth of fibrous tissue **2** the natural hardening of plant cell walls usu by the formation of lignin

sclerotic[1] *adj* **1** being or relating to the sclera **2** of or affected with sclerosis

sclerotic[2] *n* the sclera

sclerotium *n, pl* **sclerotia** a compact mass of hardened fungal mycelium that becomes detached and remains dormant until at a favourable opportunity for growth occurs

scolex *n, pl* **scolices** the head of a tapeworm

scorbutic *adj* of, resembling, or diseased with scurvy

scorpioid *adj* curved at the end like a scorpion's tail <*a ~ inflorescence*>

scorpion *n* any of an order of arachnids having an elongated body and a narrow tail bearing a venomous sting at the tip ☞ EVOLUTION

scotoma *n, pl* **scotomas, scotomata** a blind or dark spot in the visual field

scotopic *adj* relating to or being vision in dim light with eyes adapted to the dark — compare PHOTOPIC

scour *n* diarrhoea or dysentery, esp in cattle — usu pl with sing. meaning but sing. or pl in constr

scrapie *n* a usu fatal virus disease of sheep characterized by twitching, intense itching, emaciation, and finally paralysis

scrofula *n* tuberculosis of lymph glands, esp in the neck

scrofulous *adj* of or affected (as if) with scrofula or a similar disease

scrotum *n, pl* **scrota scrotums** the external pouch of most male mammals that contains the testes ☞ REPRODUCTION

scrub *n* **1** (an area covered with) vegetation consisting chiefly of stunted trees or shrubs <*~ land*> <*~ vegetation*> **2** a usu inferior type of domestic animal of mixed or unknown parentage; a mongrel

scurvy *n* a deficiency disease caused by a lack of vitamin C and marked by spongy gums, loosening of the teeth, and bleeding under the skin

scut *n* a short erect tail (eg of a hare)

scute *n* an external hard plate or large small scale (eg on the belly of a snake)

scutellum *n*, *pl* **scutella** any of several small usu hard (shield-shaped) plates or scales on a plant or animal (eg on the feet of a bird)

scutum *n*, *pl* **scuta** a scute

sea gooseberry *n* a ctenophore

sea squirt *n* any of various tunicate sea animals that are permanently attached to a surface for all their adult lives

sea urchin *n* any of a class of echinoderms usu with a thin shell covered with movable spines

seaweed *n* (an abundant growth of) a plant, specif an alga, growing in the sea, typically having thick slimy fronds

sebaceous *adj* of, secreting, or being sebum or other fatty material ☞ SKIN

sebum *n* fatty lubricant matter secreted by sebaceous glands of the skin

secondary[1] *adj* **1** not first in order of occurrence or development **2** produced away from a growing point by the activity of plant formative tissue, esp cambium *<~ growth> <~ phloem> <~ thickening>* **3** of or being the (feathers growing on the) second segment of the wing of a bird

secondary[2] *n* a secondary feather ☞ BIRD

secondary sex characteristic *n* a physical or mental attribute characteristic of a particular sex (eg the breasts of a female mammal) that appears at puberty or in the breeding season, and is not directly concerned with reproduction

secondary thickening *n* extra vascular tissue formed by the activity of the cambium in older plant stems

second-degree burn *n* a burn characterized by blistering and surface destruction of the skin — compare FIRST-DEGREE BURN, THIRD-DEGREE BURN

secrete *vt* to form and give off (a secretion)

secretion *n* (a product formed by) the bodily process of making and releasing some material either functionally specialized (eg a hormone, saliva, latex, or resin) or isolated for excretion (eg urine)

section *n* a very thin slice (eg of tissue) suitable for microscopic examination

sedative *n or adj* (sthg, esp a drug) tending to calm or to tranquillize nervousness or excitement

sedentary *adj* **1** *esp of birds* not migratory **2** permanently attached *<~ barnacles>*

seed[1] *n*, *pl* **seeds**, *esp collectively* **seed 1a** the grains or ripened ovules of plants used for sowing **b** the fertilized ripened ovule of a (flowering) plant that contains an embryo and is capable of germination to produce a new plant **2** semen **3** the condition or stage of bearing seed <*in* ~>

seed [superscript 2] *vt* **1** to plant seeds in; [superscript 2]SOW 1 <~ *land to grass*> **2** PLANT 1a

seed leaf *n* COTYLEDON 2

seed vessel *n* a pericarp

segmentation *n* the process of dividing into segments; *esp* the formation of many cells from a single cell (eg in a developing egg)

segmented *adj* divided into or composed of sections <~ *worms*>

segregate *vi* to undergo (genetic) segregation

segregation *n* the separation of pairs of genes controlling the same hereditary characteristic, that occurs during meiotic cell division

seizure *n* a sudden attack (eg of disease)

selachian *n* any of a group of cartilaginous fishes usu considered to include the sharks and dogfishes and sometimes the rays

selection *n* a natural or artificially imposed process that results in the survival and propagation only of organisms with desired or suitable attributes so that their heritable characteristics only are perpetuated in succeeding generations — compare NATURAL SELECTION

self *n, pl* **selves** the body, emotions, thoughts, sensations, etc that constitute the individuality and identity of a person

self-coloured *adj* of a single colour <*a* ~ *flower*>

self-concept *n* a self-image

self-expression *n* the expression of one's individual characteristics (eg through painting or poetry)

self-fertilization *n* fertilization by the union of ova with pollen or sperm from the same individual — compare CROSS-FERTILIZATION

self-image *n* one's conception of oneself or of one's role

self-pollination *n* the transfer of pollen from the anther of a flower to the stigma of the same or a genetically identical flower — compare CROSS-POLLINATION

self-sow *vi* **self-sown, self-sowed** *of a plant* to grow from seeds spread naturally (eg by wind or water)

self-sterile *adj* unable to produce progeny by self-fertilization

sematic *adj, of a poisonous or unpleasant animal's (bright) colours* warning of danger <*the* ~ *coloration of the skunk*>

semen *n* a suspension of spermatozoa produced by the male reproductive glands that is conveyed to the female reproductive tract during coitus

semicircular canal *n* any of the 3 loop-shaped tubular parts of the inner ear that together constitute a sensory organ associated with the maintenance of bodily equilibrium
☞ SENSE ORGAN

semiconscious *adj* not fully aware or responsive

semilunar *adj* crescent-shaped

semilunar valve *n* (any of the crescent-shaped cusps that occur in) the aortic valve or the pulmonary valve

seminal *adj* (consisting) of, storing, or conveying seed or semen <~ *duct*><~ *vesicle*> ⇨ REPRODUCTION

seminiferous *adj* producing or bearing seed or semen

semipermeable *adj, esp of a membrane* permeable to small molecules but not to larger ones

senescence *n* being or becoming old or withered

senile *adj* of, exhibiting, or characteristic of (the mental or physical weakness associated with) old age

sensation *n* 1 a mental process (eg seeing or hearing) resulting from stimulation of a sense organ 2 a state of awareness of a usu specified type resulting from internal bodily conditions or external factors; a feeling or sense <~s of *fatigue*>

sense¹ *n* 1 (the faculty of perceiving the external world or internal bodily conditions by means of) any of the senses of feeling, hearing, sight, smell, taste, etc 2 soundness of mind or judgment — usu pl with sing. meaning <*when he came to his ~s he was shocked to hear what he had done*>

sense² *vt* to perceive by the senses

sense organ *n* a bodily structure that responds to a stimulus (eg heat or sound waves) by initiating impulses in nerves that convey them to the central nervous system where they are interpreted as sensations ⊚ ⇨ SKIN

sensibility *n* ability to have sensations <*tactile* ~>

sensible *adj* capable of sensing <~ *to pain*>

sensitive *adj* 1 capable of being stimulated or excited by external agents (eg light, gravity, or contact) 2 hypersensitive <~ *to egg protein*>

sensit·ize, -ise *vb* to make or become sensitive or hypersensitive

sensorium *n, pl* **sensoriums, sensoria** (the parts of the brain or the mind concerned with the reception and interpretation of stimuli from) all the sensory apparatus

sensory *adj* of sensation or the senses ⇨ SENSE ORGAN

sepal *n* any of the modified leaves comprising the calyx of a flower ⇨ FLOWER

sepsis *n, pl* **sepses** the spread of bacteria from a focus of infection; *esp* septicaemia

septate *adj* divided by or having a septum

septic *adj* 1 putrefactive 2 relating to, involving, or characteristic of sepsis

septicaemia *n* invasion of the

Ear

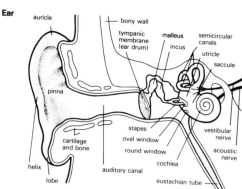

Eye

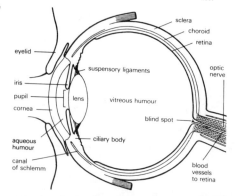

Nose

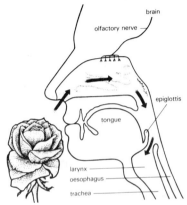

brain

olfactory nerve

epiglottis

tongue

larynx

oesophagus

trachea

Tongue

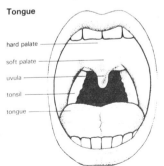

hard palate

soft palate

uvula

tonsil

tongue

Taste areas of the tongue

BITTER

SOUR   SOUR

SALT   SALT

SWEET

Sense organs relay the sensations they receive to the brain where these impulses are translated into consciousness of the world around us.

sep

bloodstream by microorganisms from a focus of infection with chills, fever, etc

septum *n, pl* **septa** a dividing wall or membrane, esp between bodily spaces or masses of soft tissue

sequence[1] *n* the order of amino acids in a protein, nucleotide bases in DNA or RNA, etc

sequence[2] *vt* to determine the amino acid sequence of (a protein), nucleotide sequence of (a nucleic acid), etc

sera *pl of* SERUM

sere *n* a series of successive ecological communities established in 1 area

serine *n* an amino acid that occurs in most proteins

serology *n* the medical study of the reactions and properties of (blood) serum

serosa *n* SEROUS MEMBRANE

serotonin *n* an amine that causes constriction of small blood vessels and occurs esp in blood platelets and as a neurotransmitter in the brain

serous membrane *n* a thin membrane (eg the peritoneum) with cells that secrete a watery liquid

serrate *adj* notched or having (forwards-pointing) teeth on the edge <*a ~ leaf*> ⫸ PLANT

serration *n* any of the teeth of a serrated edge

serum *n, pl* **serums, sera** the watery part of an animal liquid (remaining after coagulation); *esp* blood serum, esp when containing specific antibodies

sessile *adj* 1 attached directly by the base without a stalk <*a ~ leaf*> 2 permanently attached or established and not free to move about <*~ polyps*>

set[1] *vb* -tt-; **set** *vt* 1 to restore to normal position or connection after dislocation or fracturing <*~ a broken bone*> 2 to cause (eg fruit) to develop ~ *vi* 1 *of a plant part* to undergo development, usu as a result of pollination 2 *of a broken bone* to become whole by knitting together

set[2] *n* 1a a mental inclination, tendency, or habit b predisposition to act in a certain way in response to an anticipated stimulus or situation 2a a young plant or rooted cutting ready for transplanting b a small bulb, corm, or (piece of) tuber used for propagation <*onion ~s*>

seta *n, pl* **setae** a slender bristle or similar part of an animal or plant

sex *n* 1 either of 2 divisions of organisms distinguished as male or female 2 the structural, functional, and behavioural characteristics that are involved in reproduction and that distinguish males and females

sex chromosome *n* a chromosome concerned directly with the inheritance of male or female sex

sex-linked *adj* (determined by a gene) located in a sex chromosome <*a ~ gene*> <*a ~ characteristic*>

sexology *n* the study of (human) sexual behaviour

sexual *adj* **1** of or associated with sex or the sexes <*~ conflict*> **2** having or involving sex <*~ reproduction*>

sexual intercourse *n* intercourse with genital contact **a** involving penetration of the vagina by the penis; coitus **b** other than penetration of the vagina by the penis

shaft *n* **1** the trunk of a tree **2** the central stem of a feather

shank *n* **1** a leg; *specif* the part of the leg between the knee and the ankle in human beings or the corresponding part in various other vertebrates **2** the stem or stalk of a plant

sheath *n, pl* **sheaths** a cover or case of a (part of a) plant or animal body <*the leaves of grasses form a ~ round the main stalk*>

sheathe *vt* to withdraw (a claw) into a sheath

shed *vb* **-dd-**; **shed** *vt* to cast off or let fall (a natural covering) ~ *vi* to cast off hairs, threads etc; moult <*the dog is ~ding*>

shell¹ *n* **1a** a hard rigid often largely calcium-containing covering of an animal (eg a turtle, oyster, or beetle) **b** a seashell **c** the hard or tough outer covering of an egg, esp a bird's egg **2** the covering or

outside part of a fruit or seed, esp when hard or fibrous **3** shell material or shells <*an ornament made of ~*>

shell² *vt* to take out of a natural enclosing cover (eg a shell, husk, pod, or capsule) <*~ peanuts*> ~ *vi* to fall out of the pod or husk <*nuts which ~ on falling from the tree*>

shelled *adj* **1** having a shell, esp of a specified kind — often in combination <*pink-shelled*> <*thick-shelled*> **2a** having the shell removed <*~ oysters*> <*~ nuts*> **b** removed from the pod or cob <*~ peas*>

shield *n* a protective structure (eg a carapace, scale, or plate) of some animals

shigella *n, pl* **shigellae** *also* **shigellas** any of a genus of bacteria that cause dysentery in animals, esp human beings

shinbone *n* TIBIA 1

shingles *n pl but sing in constr* severe short-lasting inflammation of certain ganglia of the nerves that leave the brain and spinal cord, caused by a virus and associated with a rash of blisters and often intense neuralgic pain

shock¹ *n* **1** a state of serious depression of most bodily functions associated with reduced blood volume and pressure and caused usu by severe injuries, bleeding, or burns **2** sudden stimulation of the nerves and convulsive contraction of the

213

sho

muscles caused by the passage of electricity through the body

shock² *vt* **1** to cause to undergo a physical or nervous shock **2** to cause (eg an animal) to experience an electric shock

shock therapy *n* a treatment for some serious mental disorders that involves artificially inducing a coma or convulsions

shock treatment *n* SHOCK THERAPY

shoot¹ *vb* **shot** *vt* to put forth in growing — usu + *out* — *vi* to grow or sprout (as if) by putting forth shoots

shoot² *n* **1** a stem or branch with its leaves, buds, etc, esp when not yet mature **2** an offshoot

short-lived *adj* not living or lasting long

short sight *n* myopia

shortsighted *adj* able to see near objects more clearly than distant objects; myopic

shoulder *n* **1** the part of the human body formed of bones, joints, and muscles that connects the arm to the trunk **2** a corresponding part of a lower vertebrate

shoulder blade *n* the scapula

shrimp *n, pl* **shrimps**, *esp collectively* **shrimp** any of a suborder of numerous mostly small marine 10-legged crustacean animals with a long slender body, compressed abdomen, and long legs

shrub *n* a low-growing usu several-stemmed woody plant

shunt *n* a means or mechanism for turning or thrusting aside: eg a surgical passage created between 2 blood vessels to divert blood from one part to another

Siamese twin *n* either of a pair of congenitally joined twins

sib *n* a brother or sister considered irrespective of sex; *broadly* any plant or animal of a group sharing a degree of genetic relationship corresponding to that of human sibs

sibling *n* a sib; *also* any of 2 or more individuals having 1 parent in common

sickle cell *n* an abnormal red blood cell of crescent shape that occurs in the blood of people affected with sickle-cell anaemia

sickle-cell anaemia *n* a hereditary anaemia occurring primarily in Negroes, in which the sickling of most of the red blood cells causes recurrent short periods of fever and pain

side effect *n* a secondary and usu adverse effect (eg of a drug) <*forced to stop taking the drug because of the* ~s>

sieve cell *n* an elongated tapering cell that is present in the phloem of conifers and lower vascular plants and is important in the conduction of nutrients through the plant

sieve tube *n* a tube consisting of an end-to-end series of thin-walled living cells that is present in plant phloem and is held to function chiefly in the conduction of nutrient solutions of organic compounds (eg sugars)

sight *n* **1** the process, power, or function of seeing; *specif* the one of the 5 basic physical senses by which light received by the eye is interpreted by the brain as a representation of the forms, brightness, and colour of the objects of the real world **2** a perception of an object by the eye **3** the range of vision

significance *n* the quality of being statistically significant

significant *adj* probably caused by sthg other than chance <*statistically ~ correlation between vitamin deficiency and disease*>

silage *n* fodder converted, esp in a silo, into succulent feed for livestock

silicosis *n* a disease of the lungs marked by hardening of the tissue and shortness of breath and caused by prolonged inhalation of silica dusts

siliqua , silique *n* a long narrow seed capsule that is characteristic of plants of the crucifer family

silk *n* **1** a fine continuous protein fibre produced by various insect larvae, usu for cocoons; *esp* a lustrous tough elastic fibre produced by silkworms and used for textiles **2** a silky material or filament (eg that produced by a spider)

silk gland *n* a gland (eg in an insect larva or spider) that produces a sticky fluid that is extruded in filaments and hardens into silk on exposure to air

silky *adj* having or covered with fine soft hairs, plumes, or scales

silo *n, pl* **silos** a trench, pit, or esp a tall cylinder (eg of wood or concrete) usu sealed to exclude air and used for making and storing silage

silt[1] *n* **1** loose sedimentary material particles less than 1/20mm in diameter **2** a deposit of sediment (eg at the bottom of a river)

silt[2] *vb* to make or become choked or obstructed with silt — often + up

Silurian *adj* of or being the period of the Palaeozoic era between the Ordovician and Devonian

simian *adj or n* (of or resembling) a monkey or ape

simple *adj* **1** free of secondary complications <*a ~ fracture*> **2** not made up of many like units <*a ~ eye*> **3** not subdivided into branches or leaflets **4** consisting of a single carpel **5** *of a fruit* developing from a single ovary

sinew *n* a tendon; *also* one prepared for use as a cord or thread

single *adj, of a plant or flower* having the normal number of petals or ray flowers — compare DOUBLE

single-blind *adj* of or being an experimental procedure which is designed to eliminate false results, in which the experimenters, but not the subjects, know the make-up of the test and control groups during the actual course of the experiments — compare DOUBLE-BLIND

sinuate *adj*, having a wavy edge with strong indentations ⮕ PLANT

sinus *n* a cavity, hollow: eg **a** a narrow passage by which pus is discharged from a deep abscess or boil **b(1)** any of several cavities in the skull that usu communicate with the nostrils and contain air **(2)** a channel for blood from the veins **(3)** a wider part in a body duct or tube (eg a blood vessel) **c** a cleft or indentation between adjoining lobes (eg of a leaf)

sinusitis *n* inflammation of a nasal sinus

sinus venosus *n* an enlarged pouch that adjoins the heart and is the passage through which blood from the veins enters the heart in lower vertebrates and in the embryos of higher vertebrates

siphon, syphon *n* any of various tubular organs in animals, esp molluscs or arthropods

siphonophore *n* any of an order of transparent free-swimming or floating marine invertebrate animals that live as colonies

sire¹ *n* the male parent of a (domestic) animal

sire² *vt* to beget — esp with reference to a male domestic animal

sirenian *n* any of an order of aquatic plant-eating mammals including the manatee and dugong ⮕ MAMMAL

skeletal *adj* of, forming, attached to, or resembling a skeleton

skeleton *n* a supportive or protective usu rigid structure or framework of an organism; *esp* the bony or more or less cartilaginous framework supporting the soft tissues and protecting the internal organs of a vertebrate ⮕ ANATOMY

skew¹ to distort from a true value or symmetrical curve <∼ed *statistical data*>

skew² *adj* more developed on one side or in one direction than another; not symmetrical

skin *n* **1** the external covering of an animal (eg a fur-bearing mammal or a bird) separated from the body, usu with its hair or feathers; pelt **2** the external limiting layer of an animal body, esp when forming a tough but flexible cover ● **3** any of various outer or surface layers (eg a rind, husk, or film)

skin graft *n* a piece of skin that is taken from one area to replace skin in a defective or damaged area

skull *n* the skeleton of the head of a vertebrate animal forming a bony or cartilaginous case that encloses and protects the brain and chief sense organs and supports the jaws

sleep[1] *n* **1** the natural periodic suspension of consciousness that is essential for the physical and mental well-being of higher animals **2a** a state marked by a diminution of feeling followed by tingling <*his foot went to* ~> **b** the state of an animal during hibernation

sleep[2] *vi* **slept** to rest in a state of sleep

sleeping sickness *n* a serious disease that is prevalent in much of tropical Africa, is marked by fever and protracted lethargy, and is caused by either of 2 trypanosomes and transmitted by tsetse flies

sleepwalker *n* a somnambulist

slime mould *n* any of a group of living organisms usu held to be lower fungi that consist of a mobile mass of fused cells and reproduce by spores

slip *n* a small shoot or twig cut for planting or grafting; a scion

slipped disc *n* a protrusion of 1 of the cartilage discs that normally separate the spinal vertebrae, producing pressure on spinal nerves and usu resulting in intense pain, esp in the region of the lower back

slough[1] *n* a place of deep mud or mire

slough[2] *also* **sluff** *n* **1** the cast-off skin of a snake **2** a mass of dead tissue separating from an ulcer

slough[3] *also* **sluff** *vi* **1** to become

shed or cast off **2** to separate in the form of dead tissue from living tissue

small intestine *n* the part of the intestine that lies between the stomach and colon, consists of duodenum, jejunum, and ileum, secretes digestive enzymes, and is the chief site of the absorption of digested nutrients

smallpox *n* an acute infectious feverish virus disease characterized by skin eruption with pustules, sloughing, and scar formation

smear *n* material smeared on a surface; *also* material taken or prepared for microscopic examination by smearing on a slide <*a vaginal* ~>

smell[1] *vb* **smelled, smelt** *vt* to perceive the odour of (as if) by use of the sense of smell ~ *vi* to exercise the sense of smell

smell[2] *n* **1a** the process, function, or power of smelling **b** the one of the 5 basic physical senses by which the qualities of gaseous or volatile substances in contact with certain sensitive areas in the nose are interpreted by the brain as characteristic odours **2** an odour **3** an act or instance of smelling

smooth muscle *n* muscle that consists of fibres usu bound in thin sheets, is present in the walls of the gut, bladder, blood vessels, etc, and is not under voluntary control — compare STRIATED MUSCLE

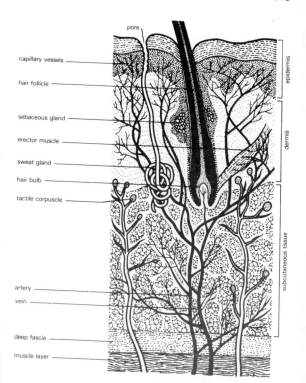

- pore
- capillary vessels
- hair follicle
- sebaceous gland
- erector muscle
- sweat gland
- hair bulb
- tactile corpuscle
- artery
- vein
- deep fascia
- muscle layer
- epidermis
- dermis
- subcutaneous tissue

smut *n* any of various destructive fungous diseases, esp of cereal grasses, marked by transformation of plant organs into dark masses of spores

snake *n* any of a suborder of limbless scaly reptiles with a long tapering body and with salivary glands often modified to produce venom which is injected through grooved or tubular fangs

social *adj* 1 living and breeding in more or less organized communities <~ *insects*> 2 *of a plant* tending to grow in patches or clumps so as to form a pure stand

society *n* 1 a voluntary association of individuals for common ends 2 a natural group of plants, usu of a single species or habit

sociobiology *n* the scientific study of animal behaviour from the point of view that all behaviour has evolved by natural selection

socket *n* an opening or hollow that forms a holder for sthg <*the eye ~*>

sodium pump *n* the process by which sodium ions are actively transported across a cell membrane

soft palate *n* the fold at the back of the hard palate that partially separates the mouth and pharynx ☞ SENSE ORGAN

softwood *n* the wood of a coniferous tree

soil[1] *n* 1 the upper layer of earth that may be dug or ploughed and in which plants grow 2 the superficial unconsolidated and usu weathered part of the mantle of a planet, esp the earth

soil science *n* the scientific study of soils

sol *n* a fluid colloidal system

solar plexus *n* 1 an interlacing network of nerves in the abdomen behind the stomach 2 the pit of the stomach

soldier *n* any of a caste of ants or wingless termites having a large head and jaws

sole *n* the undersurface of a foot

solifluction *n* the slow creeping, esp of saturated soil, down a slope that usu occurs in regions of perennial frost

solitary *adj* growing or living alone; not gregarious, colonial, social, or compound

somatic *adj* 1 of or affecting the body, esp as distinguished from the germ cells or the mind 2 of the wall of the body; parietal

somatic cell *n* any of the cells of the body that compose its tissues, organs, and other parts other than the germ cells

somatoplasm *n* somatic cells as distinguished from germ cells

somatotrophic hormone *n* GROWTH HORMONE 1

somatotype *n* body type; physique

somite *n* any of the longitudinal series of body segments of a higher invertebrate or embryonic vertebrate

somnambulist *n* sby who walks in his/her sleep

somnolent *adj* inclined to or heavy with sleep

songbird *n* **1** a bird that utters a succession of musical tones **2** a passerine bird

sonic *adj, of waves and vibrations* having a frequency within the audibility range of the human ear

soporific *adj* **1** causing or tending to cause sleep **2** of or marked by sleepiness or lethargy

sore *n* a localized sore spot on the body; *esp* one (eg an ulcer) with the tissues ruptured or abraded and usu infected

sorghum *n* any of an economically important genus of Old World tropical grasses similar to maize in habit but with the spikelets in pairs on a hairy stalk

sorus *n, pl* **sori** a cluster of plant reproductive bodies of a lower plant; *esp* any of the dots on the underside of a fertile fern frond consisting of a cluster of spores

sound¹ *adj* healthy

sound² *n* **1** the sensation perceived by the sense of hearing **2** a particular auditory impression or quality *<the ~ of children playing>*

sound³ *vi* to make a sound ~ *vt* to examine by causing to emit sounds

sound⁴ *n* the air bladder of a fish

sound⁵ *vt* to explore or examine (a body cavity) with sound ~ *vi, of a fish or whale* to dive down suddenly

sound⁶ *n* a probe for exploring or sounding body cavities

sour¹ *adj* **1** being or inducing the one of the 4 basic taste sensations that is produced chiefly by acids *<~ pickles>* — compare BITTER, SALT, SWEET ☞ SENSE ORGAN **2a** having the acid taste or smell (as if) of fermentation *<~ cream>* **b** of or relating to fermentation **3** *esp of soil* acid in reaction

sour² *n* the primary taste sensation produced by sthg sour

sour³ *vb* to make or become sour

sow *vb* **sowed; sown, sowed** *vi* to plant seed for growth, esp by scattering ~ *vt* **1** to scatter (eg seed) on the earth for growth; *broadly* PLANT **1a 2** to strew (as if) with seed **3** to introduce into a selected environment

soya bean *n* (the edible oil-rich and protein-rich seeds of) an annual Asiatic leguminous plant widely grown for its seed and soil improvement

spadix *n, pl* **spadices** a spike of crowded flowers (eg in an arum) with a fleshy or succulent axis usu enclosed in a spathe

spasm *n* an involuntary and abnormal muscular contraction

spastic¹ *adj* **1** of or characterized by spasm *<a ~ colon>* **2** suffering from spastic paralysis *<a ~ child>*

spastic[2] *n* one who is suffering from spastic paralysis

spastic paralysis *n* paralysis with involuntary contraction or uncontrolled movements of the affected muscles — compare CEREBRAL PALSY

spathe *n* a sheathing bract or pair of bracts enclosing the inflorescence of a plant, esp a spadix on the same axis <the ~ of cuckoopint>

spatulate *adj* shaped like a spatula <~ spines of a caterpillar> ⊃ PLANT

spawn[1] *vt* **1** *of an aquatic animal* to produce or deposit (eggs) **2** to bring forth, esp abundantly ~ *vi* **1** to deposit spawn **2** to produce young, esp in large numbers

spawn[2] *n* **1** the large number of eggs of frogs, oysters, fish, etc **2** *sing or pl in constr* (numerous) offspring **3** mycelium, esp for propagating mushrooms

spear *n* a usu young blade, shoot, or sprout (eg of asparagus or grass)

special·ize, -ise *vi* to undergo structural adaptation of a body part to a particular function or of an organism for life in a particular environment

species *n, pl* **species 1** a category in the biological classification of living things that ranks immediately below a genus, comprises related organisms or populations potentially capable of interbreeding, and is designated by a name (eg *Homo*

sapiens) that consists of the name of a genus followed by a Latin or latinized uncapitalized noun or adjective **2** an individual or kind belonging to a biological species

specific *adj* **1** having a specific rather than a general influence (eg on a body part or a disease) <*antibodies ~ for the smallpox virus*> **2** of or constituting a (biological) species

speculum *n, pl* **specula** *also* **speculums 1** an instrument inserted into a body passage for medical inspection or treatment **2** a patch of colour on the secondary feathers of many birds, esp ducks

sperm *n, pl* **sperms**, *esp collectively* **sperm 1** the male fertilizing fluid; semen **2** a male gamete

spermary *n* an organ in which male gametes are developed

spermatheca *n* a sac for sperm storage in the female reproductive tract of many lower animals

spermatic *adj* relating to, resembling, carrying, or full of sperm

spermatic cord *n* a cord that suspends the testis within the scrotum

spermatid *n* any of the cells that form spermatozoa

spermatium *n, pl* **spermatia** a nonmotile cell functioning or held to function as a male gamete in some lower plants

spermatocyte *n* a cell giving rise to sperm cells; *esp* a cell of the (next to) last generation preceding the spermatozoon

spermatogenesis *n* the process of male gamete formation including meiotic cell division and transformation of the 4 resulting spermatids into spermatozoa

spermatogonium *n, pl* **spermatogonia** a primitive male germ cell

spermatophore *n* a capsule, packet, or mass enclosing spermatozoa produced by the male and conveyed to the female in the insemination of various invertebrates (eg the spider)

spermatophyte *n* any of a group of higher plants constituting those that produce seeds ⟿ PLANT

spermatozoid *n* a motile male gamete of a plant, usu produced in an antheridium

spermatozoon *n, pl* **spermatozoa** 1 a motile male gamete of an animal, usu with rounded or elongated head and a long tail-like flagellum 2 a spermatozoid

sperm cell *n* a male gamete or germ cell

spermiogenesis *n* 1 transformation of a spermatid into a spermatozoon 2 spermatogenesis

sphagnum *n* any of a large genus or order of mosses that grow only in wet acid areas (eg bogs) where their remains become compacted with other plant debris to form peat

sphincter *n* a muscular ring, surrounding and able to contract or close a bodily opening

sphygmomanometer *n* an instrument for measuring (arterial) blood pressure

spicate *adj* pointed, spiked <*a ~ inflorescence*>

spicule *n* a minute slender pointed usu hard body; *esp* any of the minute bodies composed of calcium carbonate or silica that together support the tissue of various invertebrates (eg a sponge)

spider *n* any of an order of arachnids having a body with 2 main divisions, 4 pairs of walking legs, and 2 or more pairs of abdominal spinnerets for spinning threads of silk used for cocoons, nests, or webs ⟿ EVOLUTION

spike *n* 1 an ear of grain 2 an elongated plant inflorescence with the flowers stalkless on a single main axis ⟿ FLOWER

spikelet *n* a small or secondary spike; *specif* any of the small spikes that make up the compound inflorescence of a grass or sedge

spin *vb* **-nn-**; **spun** *vi, esp of a spider or insect* to form a thread by forcing out a sticky rapidly hardening fluid ~ *vt* 1 to draw out and twist into yarns or threads 2 to form (eg a web or cocoon) by spinning

spina bifida *n* a congenital condition in which there is a defect in the formation of the spine allowing the meninges to protrude and usu associated with disorder of the nerves supplying the lower part of the body

spinal adj 1 of or situated near the backbone 2 of or affecting the spinal cord <~ *reflexes*> 3 of or resembling a spine

spinal canal n a canal that contains the spinal cord

spinal column n the skeleton running the length of the trunk and tail of a vertebrate that consists of a jointed series of vertebrae and protects the spinal cord

spinal cord n the cord of nervous tissue that extends from the brain lengthways along the back in the spinal canal, carries impulses to and from the brain, and serves as a centre for initiating and coordinating many reflex actions ➔ ANATOMY

spinal nerve n any of the paired nerves that arise from the spinal cord, supply muscles of the trunk and limbs, and normally form 31 pairs in human beings ➔ ANATOMY

spindle n a spindle-shaped figure seen in microscopic sections of dividing cells along which the chromosomes are distributed

spine n 1 SPINAL COLUMN ➔ ANATOMY 2 a stiff pointed plant part; esp one that is a modified leaf or leaf part 3 a sharp rigid part of

an animal or fish; also a pointed prominence on a bone

spinneret n an organ, esp of a spider or caterpillar, for producing threads of silk from the secretion of silk glands

spinose adj SPINY 1 <a fly with black ~ legs>

spinule n a minute spine

spiny adj 1 covered or armed with spines; broadly bearing spines, prickles, or thorns 2 slender and pointed like a spine

spiracle n a breathing orifice (eg the blowhole of a whale or a tracheal opening in an insect) ➔ INSECT

spirochaete, NAm chiefly **spirochete** n any of an order of slender spirally undulating bacteria including those causing syphilis and relapsing fever

spirograph n an instrument for recording respiratory movements

spirometer n an instrument for measuring the air entering and leaving the lungs

splanchnic adj of the viscera

spleen n a highly vascular ductless organ near the stomach or intestine of most vertebrates that is concerned with final destruction of blood cells, storage of blood, and production of lymphocytes

splenectomy n surgical removal of the spleen

splenic adj of or located in the spleen <~ *blood flow*>

splenomegaly n enlargement of the spleen

split personality n a personality composed of 2 or more internally consistent groups of behaviour tendencies and attitudes each acting more or less independently of the other

spondylitis n inflammation of the spinal vertebrae

sponge n 1 an elastic porous mass of interlacing horny fibres that forms the internal skeleton of various marine animals and is able when wetted to absorb water 2 any of a phylum of aquatic lower invertebrate animals that are essentially double-walled cell colonies and permanently attached as adults ⌁ EVOLUTION

spontaneous adj 1 controlled and directed internally 2 developing without apparent external influence, force, cause, or treatment <~ recovery from a severe illness>

spontaneous generation n abiogenesis

spoor n a track, a trail, or droppings, esp of a wild animal

sporangiophore n a stalk or receptacle bearing sporangia

sporangium n, pl **sporangia** a case or cell within which usu asexual spores are produced

spore n a primitive usu single-celled hardy reproductive body produced by plants, protozoans, bacteria, etc and capable of development into a new individual either on its own or after fusion with another spore

sporocyst n a resting cell (eg in slime moulds and algae) that may give rise to asexual spores

sporogenesis n reproduction by or formation of spores

sporogenous adj of, involving, or reproducing by sporogenesis

sporogony n reproduction by spores

sporophyll n a modified leaf bearing spores

sporophyte n (a member of) the generation of a plant exhibiting alternation of generations that bears asexual spores

sporozoan n any of a large class of strictly parasitic protozoans that have a complicated life cycle usu involving both asexual and sexual generations often in different hosts and include important pathogens (eg malaria parasites and coccidia)

sporozoite n an infectious form of some sporozoans that is a product of sporogony and initiates an asexual cycle in the new host

sport¹ vt to put forth as a sport or bud variation ~ vi to deviate or vary abruptly from type

sport² n an individual exhibiting a sudden deviation from type beyond the normal limits of individual variation

sporulation n the formation of spores; esp division into many small spores

sprain¹ n 1 a sudden or violent twist

or wrench of a joint with stretching or tearing of ligaments

sprain[2] *vt* to subject to sprain

spray *n* a usu flowering branch or shoot

spread *vb* **spread** *vt* to extend the range or incidence of <~ *a disease*> ~ *vi* to become dispersed, distributed, or scattered <*a race that ~ across the globe*>

sprig *n* a small shoot or twig

springtail *n* any of an order of small primitive wingless insects ☞ IN-SECT

springwood *n* the softer more porous portion of an annual ring of wood that develops early in the growing season

sprout[1] *vi* 1 to grow, spring up, or come forth as (if) a shoot 2 to send out shoots or new growth ~ *vt* to send forth or up; cause to develop or grow

sprout[2] *n* a (young) shoot (eg from a seed or root)

spur *n* 1 a stiff sharp spine (eg on the wings or legs of a bird or insect); *esp* one on a cock's leg 2 a hollow projection from a plant's petals or sepals (eg in larkspur or columbine)

sputum *n, pl* **sputa** matter, made up of discharges from the respiratory passages and saliva, that is coughed up

squama *n, pl* **squamae** (a structure resembling) a scale

squamous *also* **squamose** *adj* 1 covered with or consisting of scales 2 of or being a surface tissue consisting of a single layer of flat scalelike cells

squid *n, pl* **squids,** *esp collectively* **squid** any of numerous 10-armed cephalopod molluscs, related to the octopus and cuttlefish, that have a long tapered body and a tail fin on each side ☞ EVOLUTION

squint *n* (a visual disorder marked by) inability to direct both eyes to the same object because of imbalance of the muscles of the eyeball

stadium *n, pl* **stadiums** *also* **stadia** a stage in a life history; *esp* one between successive moults in the development of an insect

stage *n* 1 the small platform of a microscope on which an object is placed for examination 2 any of the distinguishable periods of growth and development of a plant or animal <*the larval ~ of an insect*>

stain[1] *vt* to colour (eg wood or a biological specimen) by using (chemical) processes or dyes affecting the material itself ~ *vi* 1 to become stained 2 to cause staining

stain[2] *n* a dye or mixture of dyes used in microscopy to make minute and transparent structures visible, to differentiate tissue elements, or to produce specific chemical reactions

stalk *n* 1a the main stem of a herbaceous plant, often with its

225

sta

attached parts **b** STEM 2 **2** a slender upright supporting or connecting (animal) structure

stalk-eyed *adj, esp of crustaceans* having the eyes raised on stalks

stamen *n* the organ of a flower that produces the male gamete in the form of pollen, and consists of an anther and a filament ➔ FLOWER

staminate *adj* **1** having or producing stamens **2** ¹MALE 2

stand *n* a group of plants or trees growing in a continuous area

standard deviation *n* a measure of the extent to which values of a variable are scattered about a mean value in a frequency distribution <*the larger the ~, the more widely dispersed are the values*>

standard error *n* the standard deviation of the distribution of values of a statistic (eg the mean) obtained from a large number of samples

stapes *n, pl* **stapes, stapedes** the innermost of the chain of 3 small bones in the ear of a mammal; the stirrup ➔ SENSE ORGAN

staphylococcus *n, pl* **staphylococci** any of various spherical bacteria that include parasites of skin and mucous membranes and cause boils, septic infections of wounds, etc

starch *n* an odourless tasteless complex carbohydrate that is the chief storage form of carbohydrate in plants, is an important foodstuff, and is used also in adhesives and in

sizes, in laundering, and in pharmacy and medicine

starchy *adj* of or containing (much) starch <~ *foods*>

starfish *n* any of a class of sea animals that are echinoderms, have a body consisting of a central disc surrounded by 5 equally spaced arms, and feed largely on molluscs (eg oysters) ➔ EVOLUTION

starter *n* **1** material containing microorganisms used to induce a desired fermentation **2** a compound used to start a chemical reaction

stasis *n, pl* **stases 1** a slowing or stoppage of the normal flow of body fluids **2** a state of static balance or equilibrium

statistic *n* a single term or quantity in or computed from a collection of statistics; *specif* (a function used to obtain) a numerical value (eg the standard deviation or mean) used in describing and analysing statistics

statistics *n pl but sing or pl in constr* **1** a branch of mathematics dealing with the collection, analysis, interpretation, and presentation of masses of numerical data **2** a collection of quantitative data

stearic acid *n* a fatty acid that is obtained from hard fat (eg tallow)

stearin *n* **1** an ester of glycerol and stearic acid **2** the solid portion of a fat

stele *n* the (cylindrical) central vascular portion of the stem of a vascular plant

stellate *also* **stellated** *adj* resembling a star, esp in shape

stelliform *adj* star-shaped

stem *n* **1** the main trunk of a plant; *specif* a primary plant axis that develops buds and shoots instead of roots **2** a branch, petiole, or other plant part that supports a leaf, fruit, etc

stemma *n, pl* **stemmata** a simple eye present in some insects

stenosis *n, pl* **stenoses** a narrowing or constriction of the diameter of a bodily passage or orifice

steppe *n* a vast usu level and treeless plain, esp in SE Europe or Asia

sterile *adj* **1** failing or not able to produce or bear fruit, crops, or offspring **2** free from living organisms, esp microorganisms

sternum *n, pl* **sternums, sterna** a bone or cartilage at the front of the body that connects the ribs, both sides of the shoulder girdle, or both; the breastbone ⇨ ANATOMY

steroid *n* any of numerous compounds of similar chemical structure, including the sterols and various hormones (eg testosterone) and glycosides

sterol *n* any of various solid alcohols (eg cholesterol) widely distributed in animal and plant fats

stethoscope *n* an instrument used to detect and study sounds produced in the body

stifle[1] *n* the joint next above the hock in the hind leg of a quadruped (eg a horse) corresponding to the knee in human beings

stifle[2] *vb* **stifling** *vt* **1** to overcome or kill by depriving of oxygen; suffocate, smother **2** to cut off (eg the voice or breath) ~*vi* to become suffocated (as if) by lack of oxygen

stigma *n, pl* **stigmata, stigmas 1** an identifying mark or characteristic; *specif* a specific diagnostic sign of a disease **2** a small spot, scar, or opening on a plant or animal **3** the portion of the female part of a flower which receives the pollen grains and on which they germinate ⇨ FLOWER

stillbirth *n* the birth of a dead infant

stimulant *n* **1** sthg (eg a drug) that produces a temporary increase in the functional activity or efficiency of (a part of) an organism **2** STIMULUS 1

stimulate *vt* **1** to excite to (greater) activity **2a** to function as a physiological stimulus to **2b** to arouse or affect by the action of a stimulant (eg a drug)

stimulus *n, pl* **stimuli 1** sthg that rouses or incites to activity; an incentive **2** sthg (eg light) that directly influences the activity of living organisms (eg by exciting a sensory organ or evoking muscular contraction or glandular secretion)

sti

sting[1] *vb* **stung** *vt* to give an irritating or poisonous wound to, esp with a sting <stung *by a bee*> ~ *vi* to use a sting; to have stings <*nettles* ~>

sting[2] *n* **1a** a stinging; *specif* the thrust of a sting into the flesh **b** a wound or pain caused (as if) by stinging **2** *also* **stinger** a sharp organ of a bee, scorpion, stingray, etc that is usu connected with a poison gland or otherwise adapted to wound by piercing and injecting a poisonous secretion

stipe *n* a usu short plant stalk (eg supporting the cap of a fungus)

stipes *n, pl* **stipites** a peduncle

stipule *n* a small appendage at the base of the leaf in many plants

stirps *n, pl* **stirpes** a race, variety, etc in the biological classification of living things

stirrup *n* the stapes

stochastic *adj* **1** random; *specif* involving a random variable <*a* ~ *process*> **2** involving chance or probability <*a* ~ *model of radiation-induced mutation*>

stock[1] *n* **1a** the main stem of a plant or tree **b** a plant (part) consisting of roots and lower trunk onto which a scion is grafted **c** a plant from which cuttings are taken **2a** the descendants of an individual; family, lineage **b** a compound organism **c** RACE 1

stock[2] *adj* used for (breeding and rearing) livestock <*a* ~ *farm*>

stolon *n* a horizontal branch from the base of a plant (eg the strawberry) that produces new plants

stoma *n, pl* **stomata** *also* **stomas 1** any of various small simple bodily openings, esp in a lower animal **2** any of the minute openings in the epidermis of a plant organ (eg a leaf) through which gases pass **3** a permanent surgically made opening, esp in the abdominal wall

stomach *n* **1** (a cavity in an invertebrate animal analogous to) a sac-like organ formed by a widening of the alimentary canal of a vertebrate, that is between the oesophagus at the top and the duodenum at the bottom and in which the first stages of digestion occur **2** the part of the body that contains the stomach; belly, abdomen

stone[1] *n, pl* **stones 1** a calculus **2** the hard central portion of a fruit (eg a peach or date)

stone[2] *vt* to remove the stones or seeds of (a fruit)

stone fruit *n* a fruit with a (large) stone; a drupe

stool[1] *n* **1** a discharge of faecal matter **2** (a shoot or growth from) a tree stump or plant crown from which shoots grow out

stool[2] *vi* to throw out shoots from a stump or crown

stoop *n* the descent of a bird, esp on its prey

STP *abbr* standard temperature and pressure

strabismus *n* a squint

strain[1] *n* a group of plants, animals, microorganisms, etc at a level lower than a species <*a high-yielding ~ of winter wheat*>

strain[2] *vt* **1** to exert (eg oneself) to the utmost **2** to injure by overuse, misuse, or excessive pressure <*~ed a muscle*> ~ *vi* **1** to sustain a strain, wrench, or distortion **2** to contract the muscles forcefully in physical exertion

strain[3] *n* **1** (a force, influence, or factor causing) physical or mental tension **2** a wrench, twist, or similar bodily injury resulting esp from excessive stretching of muscles or ligaments

strath *n* a flat wide river valley, esp in Scotland

stratum *n, pl* **strata** a layer of tissue

streak *n* a sample containing microorganisms (eg bacteria) implanted in a line on a solid culture medium (eg agar jelly) for growth

streptococcus *n, pl* **streptococci** any of a genus of chiefly parasitic bacteria that occur in pairs or chains and include some that cause diseases in human beings and domestic animals

streptomycin *n* an antibiotic obtained from a soil bacterium and used esp in the treatment of tuberculosis

stress *n* (a physical or emotional factor that causes) bodily or mental tension

striated muscle *n* muscle that is marked by alternate light and dark bands, is made up of long fibres, and comprises the voluntary muscle of vertebrates — compare SMOOTH MUSCLE

striation *n* **1a** being striated **b** an arrangement of striae **2** a stria

stricture *n* an abnormal narrowing of a bodily passage

stridulate *vi, esp of crickets, grasshoppers, etc* to make a shrill creaking noise by rubbing together special bodily structures

strigose *adj* **1** having bristles or scales lying against a surface <*a ~ leaf*> **2** marked with fine grooves <*the ~ wing cases of a beetle*>

strike *vb* **struck** *also* **stricken** *vt* **1** to afflict suddenly <*stricken by a heart attack*> **2** to send down or out <*trees struck roots deep into the soil*> **3a** to place (a plant cutting) in a medium for growth and rooting **b** to propagate (a plant) in this manner ~ *vi, of a plant cutting* to take root

strip-cropping *n* the growing of a cultivated crop (eg maize) in alternate strips with a turf-forming crop (eg hay) to minimize erosion of the land

strip farming *n* **1** the growing of crops in separate strips of land allotted to individual farmers so

str

that good and bad land is fairly distributed **2** strip-cropping

strobila *n, pl* **strobilae** a line of similar joined animal structures (eg the segmented body of a tapeworm) produced by budding

strobilation *n* the production of strobilae by asexual reproduction

strobilus *n, pl* **strobili** CONE 1

stroke *n* (an attack of) sudden usu complete loss of consciousness, sensation, and voluntary motion caused by rupture, thrombosis, etc of a brain artery

stroma *n, pl* **stromata 1** the supporting framework of an animal organ or of some cells **2a** a compact mass of fungal hyphae producing a fruiting body **b** the colourless matrix of a chloroplast in which the chlorophyll-containing layers are embedded

structural *adj* of the physical make-up of a plant or animal body

structural gene *n* the part of an operon that codes for the sequence of amino acids in a protein molecule

structure *n* the arrangement of particles or parts in a substance or body <*soil* ~><*molecular* ~>

struggle for existence *n* the competition for food, space, etc that tends to eliminate less efficient individuals of a population, thereby increasing the chance of inherited traits being passed on from the more efficient survivors — compare NATURAL SELECTION

stupor *n* a state of extreme apathy, torpor, or reduced sense of feeling (eg resulting from shock or intoxication)

sty, stye *n, pl* **sties, styes** an inflamed swelling of a sebaceous gland at the margin of an eyelid

style *n* **1** a prolongation of a plant ovary bearing a stigma at the top ⤳ FLOWER **2** a slender elongated part (eg a bristle) on an animal

stylet *n* **1a** a slender surgical probe **b** a thin wire inserted into a catheter to maintain rigidity or into a hollow needle to keep it clear of obstruction **2** a relatively rigid elongated organ or part (eg a piercing mouthpart) of an animal

subalpine *adj* of or growing on high upland slopes

subantarctic *adj* (characteristic) of or being a region just outside the antarctic circle

subaquatic *adj* somewhat aquatic

subaqueous *adj* taking place or existing under water

subarctic *adj* (characteristic) of or being a region just outside the arctic circle

subcellular *adj* occurring inside cells; *also* derived from the artificial disruption of cells <~ *particles*>

subclass *n* a category in the biological classification of living things below a class and above an order

subclavian *adj* (of or being an artery, nerve, etc) situated under the clavicle

subclinical *adj* having (practically) undetectable symptoms <*a ~ infection*>

subconscious[1] *adj* **1** existing in the mind but not immediately available to consciousness <*his ~ motive*> **2** imperfectly or incompletely conscious <*a ~ state*>

subconscious[2] *n* the mental activities below the threshold of consciousness

subculture *n* a culture (eg of bacteria) derived from another culture

subcutaneous *adj* being, living, used, or made under the skin <*~ fat*>

suberin *n* a complex waxy substance that is the basis of cork

suber·ization, -isation *n* conversion of plant cell walls into corky tissue by impregnation with suberin

subjective *adj* arising from conditions within the brain or sense organs and not directly caused by external stimuli <*~ sensations*>

sublimate *vt* to divert the expression of (an instinctual desire or impulse) from a primitive form to a socially or culturally acceptable one

subliminal *adj* **1** *of a stimulus* inadequate to produce a sensation or perception **2** existing, functioning, or having effects below the level of conscious awareness <*the ~ mind*><*~ advertising*>

sublittoral *n* the region in the sea between the lowest point exposed by a very low tide and the margin of the continental shelf

submaxilla *n, pl* **submaxillae** *also* **submaxillas** (the bone of) the lower jaw, specif in humans

submerged *adj* submersed

submersed *adj* **1** covered with water **2** (adapted for) growing under water <*~ plants*>

submicroscopic *adj* too small to be seen in an ordinary light microscope

submucosa *n* a supporting layer of loose connective tissue directly under a mucous membrane

subnormal *adj* having less of sthg, esp intelligence, than is normal

subsoil *n* the layer of weathered material that underlies the surface soil

subspecies *n* a category in the biological classification of living things that ranks (immediately) below a species

substrate *n* **1** a substratum **2** the base on which an organism lives <*limpets live on a rocky ~*> **3** a substance acted on (eg by an enzyme)

substratum *n, pl* **substrata** an underlying support; a foundation: eg the subsoil

subtend *vt* to be lower than, esp so

sub

as to embrace or enclose <*a bract that ~s a flower*>

subtropical *also* **subtropic** *adj* of or being the regions bordering on the tropical zone

subulate *adj* narrow and tapering to a fine point <*a ~ leaf*> ⇨ PLANT

succession *n* the change in the composition of an ecological system as the competing organisms respond to and modify the environment

succinic acid *n* a carboxylic acid found widely in nature and active in the Krebs cycle

succulent *adj*, *of a plant* having juicy fleshy tissues

sucker[1] *n* **1a** a human infant or young animal that sucks, esp at a breast or udder; a suckling **b** a mouth (eg of a leech) or other animal organ adapted for sucking or sticking **2** a shoot from the roots or lower part of the stem of a plant

sucker[2] *vt* to remove suckers from <*~ tobacco*> ~ *vi* to send out suckers

suckle *vt* **suckling 1** to give milk to from the breast or udder <*a mother suckling her child*> **2** to draw milk from the breast or udder of <*lambs suckling the ewes*>

suckling *n* a young unweaned animal

sucrose *n* the disaccharide sugar obtained from sugarcane and sugar beet and occurring in most plants

sudoriferous *adj* producing or conveying sweat <*~ glands*>

sugar *n* **1** a sweet crystallizable material that consists (essentially) of sucrose, is colourless or white when pure tending to brown when less refined, is obtained commercially esp from sugarcane or sugar beet, and is important as a source of dietary carbohydrate and as a sweetener and preservative of other foods **2** any of a class of water-soluble carbohydrate compounds containing many hydroxyl groups that are of varying sweetness and include glucose, ribose, and sucrose

sugar beet *n* a white-rooted beet grown for the sugar in its root

sugarcane *n* a stout tall grass widely grown in warm regions as a source of sugar

suggestion *n* the impressing of an idea, attitude, desired action, etc on the mind of another

sulcate *adj* scored with (longitudinal) furrows <*a ~ seedpod*>

sulcus *n*, *pl* **sulci** a (shallow) furrow, esp on the surface of the brain between convolutions

sullage *n* **1** refuse, sewage **2** silt

summerwood *n* the harder less porous section of an annual ring of wood that develops late in the growing season

superego *n* the one of the 3 divisions of the mind in psychoanalytic theory that is only partly conscious, reflects social rules, and functions as a conscience to reward and punish — compare EGO, ID

superfetation n successive fertilization of 2 or more ova of different ovulations resulting in the presence of embryos of unlike ages in the same uterus

superficial adj 1 of a surface 2 not penetrating below the surface <~ wounds>

superior adj 1 of an animal or plant part situated above or at the top of another (corresponding) part 2a of a calyx attached to and apparently arising from the ovary **b** of an ovary free from and above a floral envelope (eg the calyx)

superiority complex n an exaggeratedly high opinion of oneself

superovulation n production of exceptional numbers of eggs at one time

superphosphate n a fertilizer made from insoluble mineral phosphates by treatment with sulphuric acid

suppository n a readily meltable cone or cylinder of medicated material for insertion into a bodily passage or cavity (eg the rectum)

suppress vt 1 to (deliberately) exclude a thought, feeling, etc from consciousness — compare REPRESS 1 2 to inhibit the growth or development of

suppurate vi to form or discharge pus

suprarenal adj adrenal <~ gland>

survival n 1a the condition of living or continuing <the ~ of the soul after death> **b** the continuation of life or existence <problems of ~ in arctic conditions> 2 sby or sthg that survives, esp after others of its kind have disappeared

survival of the fittest n NATURAL SELECTION

susceptible adj open, subject, or unresistant to some stimulus, influence, or agency

suspended animation n temporary suspension of the vital functions (eg in people nearly drowned)

suspensory ligament n a membrane that holds the lens of the eye in position ⟿ SENSE ORGAN

suture[1] n 1a (a strand or fibre used in) the sewing together of parts of the living body **b** a stitch made with a suture 2a the solid join between 2 bones (eg of the skull) **b** a furrow at the junction of animal or plant parts

suture[2] vt to unite, close, or secure with sutures <~ a wound>

swallow[1] vt to take through the mouth and oesophagus into the stomach ~ vi to receive sthg into the body through the mouth and oesophagus

swallow[2] n 1 an act of swallowing 2 an amount that can be swallowed at one time

swamp n (an area of) wet spongy land sometimes covered with water

swa

sward *n* (a piece of ground covered with) a surface of short grass

swarm¹ *n* **1** a colony of honeybees, esp when emigrating from a hive with a queen bee to start a new colony elsewhere **2** a cluster of free-floating or free-swimming zoospores or other single-celled organisms

swarm² *vi* to collect together and depart from a hive

swarm spore *n* a zoospore or other minute mobile spore

sweat¹ *vb* **sweated**, *NAm chiefly* **sweat** *vi* to excrete sweat in visible quantities ~ *vt* **1** to (seem to) emit from pores; exude **2** to cause (eg a patient) to sweat **3** to subject (esp tobacco) to fermentation

sweat² *n* the fluid excreted from the sweat glands of the skin; perspiration

sweat gland *n* a tubular gland in the skin that secretes sweat through a minute pore on the surface of the skin

sweet *adj* **1** being or inducing the one of the 4 basic taste sensations that is typically induced by sucrose — compare BITTER, SALT, SOUR ⟿ SENSE ORGAN **2a** not sour, rancid, decaying, or stale **b** not salt or salted; fresh <~ *butter*> <~ *water*>

swim bladder *n* the air bladder of a fish

swimmeret *n* any of a series of small unspecialized appendages under the abdomen of many crustaceans that are used for swimming or carrying eggs

swine fever *n* a highly infectious often fatal virus disease of pigs

syconium *n*, *pl* **syconia** a multiple fleshy fruit (eg a fig) in which the ovaries are borne within an enlarged succulent receptacle

sylvan, silvan *adj* **1** of, located in, or characteristic of the woods or forest **2** full of woods or trees

symbiont *n* an organism living in symbiosis

symbiosis *n*, *pl* **symbioses** the living together of 2 dissimilar organisms in intimate association (to their mutual benefit)

symbiote *n* a symbiont

symmetrical, symmetric *adj*, *of a flower* having the same number of members in each whorl of floral leaves

symmetry *n* the property of being symmetrical; *esp* correspondence in size, shape, and relative position of parts on opposite sides of a dividing line or median plane or about a centre or axis — compare BILATERAL SYMMETRY, RADIAL SYMMETRY

sympathetic *adj* **1** not discarded or antagonistic **2** of, being, mediated by, or acting on (the nerves) of the sympathetic nervous system

sympathetic nervous system *n* the part of the autonomic nervous system that contains nerve fibres in which the chief neurotransmitter is noradrenalin and whose activity tends to relax smooth muscle and cause the contraction of blood vessels — compare PARASYMPATHETIC NERVOUS SYSTEM

sympathomimetic *adj* simulating sympathetic nervous action in physiological effect <~ *drugs*>

sympetalous *adj* gamopetalous

symphysis *n, pl* **symphyses** an (almost) immovable joint between bones, esp where the surfaces are connected by fibrous cartilage without a joint membrane

sympodial *adj* having or involving the formation of an apparent main axis (eg of an inflorescence) from successive secondary axes

symptom *n* sthg giving (subjective) evidence or indication of disease or physical disturbance

synapse *n* the point (between 2 nerves) across which a nervous impulse is transmitted

synapsis *n, pl* **synapses** the joining of homologous chromosomes that occurs in meiotic cell division

syncarpous *adj, of a flower, fruit, etc* having the carpels united in a compound ovary

syncytium *n, pl* **syncytia** (an organism consisting of) a mass of living material with many nuclei resulting from fusion of cells or repeated division of nuclei

syndrome *n* 1 a group of signs and symptoms that occur together and characterize a particular (medical) abnormality 2 a set of concurrent emotions, actions, etc that usu form an identifiable pattern

synecology *n* ecology that deals with the structure and development of ecological communities

syngamy *n* sexual reproduction by union of gametes

synovia *n* a transparent viscous lubricating fluid secreted by a joint or tendon membrane

synovitis *n* inflammation of a synovial membrane

syphilis *n* a contagious usu venereal and often congenital disease caused by a spirochaetal bacterium

syringe[1] *n* a device used to inject fluids into or withdraw them from sthg (eg the body or its cavities); *esp* one that consists of a hollow barrel fitted with a plunger and a hollow needle

syringe[2] *vt* to irrigate or spray (as if) with a syringe

syrinx *n, pl* **syringes, syrinxes** the vocal organ of birds that is a modification of the lower trachea, bronchi, or both

systaltic *adj* alternately and regularly contracting and dilating; pulsating

system *n* **1a** a group of body organs that together perform 1 or more usu specified functions <the diges-tive ~> **b** the body considered as a functional unit **2** a manner of classifying, symbolizing, or formalizing <a taxonomic ~>

sys

systematic *also* **systematical** *adj* of or concerned with classification; *specif* taxonomic

systematics *n pl but sing in constr* (a system of) classification or taxonomy

systematist *n* a taxonomist

systemic *adj* **1** affecting the body generally **2** *of an insecticide, pesticide, etc* making the organism, esp a plant, toxic to a pest by entering the tissues

systemic circulation *n* the part of the blood circulation concerned with the distribution of blood to the tissues through the aorta rather than to the lungs through the pulmonary artery

systole *n* the recurrent contraction of the heart by which blood is forced on and the circulation kept up — compare DIASTOLE

tachycardia *n* normal or abnormal rapid heart action — compare BRADYCARDIA

tactile *adj* of or perceptible by (the sense of) touch

tadpole *n* the larva of an amphibian; *specif* a frog or toad larva with a rounded body, a long tail, and external gills

taenia *n, pl* **taeniae, taenias 1** a band of nervous tissue or muscle **2** any of numerous tapeworms

taiga *n* moist coniferous forest that begins where the tundra ends and is dominated by spruces and firs

taint *vt* to affect with putrefaction;

spoil ~ *vi* to become affected with putrefaction; spoil

tall *adj, of a plant* of a higher growing variety or species

talon *n* a claw of an animal, esp a bird of prey

talus *n, pl* **tali 1** the astragalus of a vertebrate, esp a human being; the anklebone **2** the ankle joint formed from the talus, the tibia, and fibula

tampon *vt or n* (to plug with) an absorbent plug put into a cavity (eg the vagina) to absorb secretions, arrest bleeding, etc

tannic acid *n* tannin

tannin *n* any of various soluble astringent complex phenolic substances of plant origin

tapeworm *n* any of numerous cestode worms, which when adult are parasitic in the intestine of human beings or other vertebrates

taproot *n* a main root of a plant that grows vertically downwards and gives off small side roots

tarsometatarsus *n* (the limb segment supported by) the large compound bone of the tarsus of a bird

tarsus *n, pl* **tarsi 1** (the small bones that support) the back part of the foot of a vertebrate that includes the ankle and heel **2** the part of the limb of an arthropod furthest from the body ⟶ INSECT **3** the plate of dense connective tissue that stiffens the eyelid

tassel *n* the tassel-like flower clusters of some plants, esp maize

taste[1] *vt* to perceive or recognize (as if) by the sense of taste <*could ~ the salt on his lips*> ~ *vi* to have a specified flavour — often + *of* <*the milk* ~s *sour*>

taste[2] *n* (the quality of a dissolved substance as perceived by) the 1 of the 5 basic physical senses by which the qualities of dissolved substances in contact with taste buds on the tongue are interpreted by the brain as 1 or a combination of the 4 basic taste sensations sweet, bitter, sour, or salt

taste bud *n* any of the small organs, esp on the surface of the tongue, that receive and transmit the sensation of taste

taxa *pl of* taxon

taxidermy *n* the art of preparing, stuffing, and mounting the skins of animals

taxis *n, pl* **taxes 1** the manual restoration of a displaced body part, esp a hernia, by pressure **2** (a reflex reaction involving) movement by a freely motile usu simple organism (eg a bacterium) towards or away from a source of stimulation (eg a light, or a temperature or chemical gradient) — compare TROPISM

taxon *n, pl* **taxa** *also* **taxons** (the name of) a taxonomic group or entity

taxonomy *n* (the study of the principles of) classification, specif of plants and animals according to

their presumed natural relationships

TB *n* tuberculosis

t distribution *n* a probability density function that is used esp in testing hypotheses concerning means of normal distributions whose standard deviations are unknown

teat *n* a nipple

ted *vt* **-dd-** to turn over and spread (eg new-mown grass) for drying

tegument *n* an integument

telencephalon *n* the front subdivision of the brain comprising the cerebral hemispheres and associated structures

teleost *n* BONY FISH

telophase *n* the final stage of cell division in which the mitotic spindle disappears and 2 new nuclei appear, each with a set of chromosomes

telson *n* the last segment of the body of an arthropod, esp a crustacean, or of a segmented worm

temperament *n* a person's peculiar or distinguishing mental or physical character (which according to medieval physiology was determined by the relative proportions of the humours)

temperate *adj* found in or associated with a temperate climate

template, templet *n* a molecule (eg of RNA) in a biological system that carries the genetic code for protein or other macromolecules

tem

temple *n* the flattened space on either side of the forehead of some mammals (eg human beings)

temporal lobe *n* a large lobe at the side of each cerebral hemisphere that contains a sensory area associated with hearing and speech

tendon *n* a tough cord or band of dense white fibrous connective tissue that connects a muscle with a bone or other part and transmits the force exerted by the muscle

tendril *n* a slender spirally coiling sensitive organ that attaches a plant to its support

tension *n* **1** the stress resulting from the elongation of an elastic body **2** inner striving, unrest, or imbalance, often with physiological indication of emotion

tensor *n* a muscle that stretches a body part

tentacle *n* **1** any of various elongated flexible animal parts, chiefly on the head or about the mouth, used for feeling, grasping, etc **2** a sensitive hair on a plant (eg the sundew)

tergum *n, pl* **terga** the plate forming the back surface of a segment of an arthropod

term *n* the time at which a pregnancy of normal length ends <*had her baby at full ~*>

terminal *adj* **1** growing at the end of a branch or stem <*a ~ bud*> **2** occurring at or causing the end of life <*~ cancer*>

ternary *adj* ternate

ternate *adj* **1** arranged in threes <*~ leaves*> **2** composed of 3 leaflets or subdivisions ⟹ PLANT

terpene *n* any of various hydrocarbons present in essential oils (eg from conifers)

terrestrial *adj, of organisms* living on or in land or soil

terricolous *adj* living on or in the ground

territorial *adj* exhibiting territoriality <*~ birds*>

territoriality *n* (the pattern of behaviour associated with) the defence of a territory

territory *n* an area, often including a nesting site or den, occupied and defended by an animal or group of animals

tertian *adj, of malarial symptoms* recurring at approximately 48-hour intervals

tertiary *adj* **1** *cap* of or being the first period of the Cainozoic era or the corresponding system of rocks **2** occurring in or being a third stage

testa *n, pl* **testae** the hard external coat of a seed

testaceous *adj* **1** having a shell **2** light brown

testcross *n* a genetic cross between a homozygous recessive individual and a corresponding suspected heterozygote to determine the genetic constitution of the latter

testicle *n* a testis, esp of a mammal

238

and usu with its enclosing structures (eg the scrotum)

testis *n, pl* **testes** a male reproductive gland ☞ REPRODUCTION

testosterone *n* a male steroid hormone, produced by the testes or made synthetically, that induces and maintains male secondary sex characters

test-tube *adj, of a baby* conceived by artificial insemination, esp outside the mother's body

tetanus *n* **1** (the bacterium, usu introduced through a wound, that causes) an infectious disease characterized by spasm of voluntary muscles, esp of the jaw **2** prolonged contraction of a muscle resulting from rapidly repeated motor impulses

tetany *n* muscle spasm usu associated with deficient secretion of parathyroid hormones

tetrad *n* a group or arrangement of 4 cells, atoms, etc

tetramerous *adj* having or characterized by (sets or multiples of) 4 parts <~ *flowers*>

tetraploid *adj* having or being a chromosome number 4 times the haploid number

tetrapod *n* a vertebrate animal with 2 pairs of limbs

thalamus *n, pl* **thalami** the subdivision of the midbrain that forms a coordinating centre through which different nerve impulses are directed to appropriate parts of the brain cortex

thalassaemia *n* a hereditary anaemia common in Mediterranean regions and characterized esp by abnormally small red blood cells

thalassic *adj* of deep seas <~ *fishes*>

thallophyte *n* any of a primary group of living things with a plant body, typically a thallus, that includes the algae, fungi, and lichens ☞ PLANT

thallus *n, pl* **thalli, thalluses** a plant body (eg of an alga) that lacks differentiation into distinct tissues or parts (eg stem or leaves)

theca *n, pl* **thecae** **1** an urn-shaped spore receptacle of a moss **2** an enveloping sheath or case of an animal (part)

therapeutic *adj* of the treatment of disease or disorders by remedial agents or methods

therapeutics *n pl but sing or pl in constr* medicine dealing with the application of remedies to diseases

therapist *n* sby trained in methods of treatment and rehabilitation other than the use of drugs or surgery <*a speech* ~>

therapy *n* therapeutic treatment of bodily, mental, or social disorders

thermocline *n* a layer of water in a lake, sea, etc that separates an upper warmer zone from a lower colder zone; *specif* a stratum in which temperature declines at least 1°C with each metre increase in depth

thermogram *n* the record made by thermography

thermography *n* a technique for photographically recording variations in the heat emitted by various regions, esp of the body (eg for the detection of tumours)

thermolabile *adj* unstable, specif losing characteristic properties, when heated above a moderate temperature

thermophile *n* a living organism thriving at relatively high temperatures

thermoregulation *n* the natural maintenance of the living body at a constant temperature

thermostable *adj* stable, specif retaining characteristic properties, when heated above a moderate temperature

thermotaxis *n* **1** a taxis in which a temperature gradient constitutes the directive factor **2** the regulation of body temperature

thermotropism *n* a tropism in which a temperature gradient is the orienting factor

thiamine *also* **thiamin** *n* a vitamin of the vitamin B complex that is essential to normal metabolism and nerve function and is widespread in plants and animals

thicket *n* a dense growth of shrubbery or small trees

thigh *n* the segment of the vertebrate hind limb nearest the body that extends from the hip to the knee and is supported by a single large bone

thighbone *n* the femur

thin-layer chromatography *n* chromatography in which the absorbent medium is a thin layer on a support (eg a glass plate)

third-degree burn *n* a burn characterized by destruction of the skin and possibly the underlying tissues, loss of fluid, and sometimes shock — compare FIRST-DEGREE BURN, SECOND-DEGREE BURN

thorax *n, pl* **thoraxes, thoraces** (a division of the body of an insect, spider, etc corresponding to) the part of the mammalian body between the neck and the abdomen; *also* its cavity in which the heart and lungs lie

thorn *n* **1** a woody plant (of the rose family) bearing sharp prickles or thorns **2** a short hard sharp-pointed plant part, specif a leafless branch

thoroughbred¹ *adj* bred from the best blood through a long line; purebred

thoroughbred² *n* a purebred or pedigree animal

thread *n* any of various natural filaments <*the ~s of a spider's web*>

threadworm *n* any of various small usu parasitic nematode worms that infest the intestines, esp the caecum, of vertebrates

thremmatology *n* the science of breeding animals and plants in domestication

threonine *n* an amino acid found in most proteins and essential to normal nutrition

threshold *n* **1** the point at which a physiological or psychological effect begins to be produced by a stimulus of increasing strength **2** a level, point, or value above which sthg is true or will take place

thrips *n*, *pl* **thrips** any of an order of small sucking insects, most of which feed on and damage plants

throat *n* **1a** the part of the neck in front of the spinal column **b** the passage through the neck to the stomach and lungs **2** the opening of a tubular (plant) organ

thrombin *n* an enzyme formed from prothrombin that acts in the process of blood clotting by catalysing the conversion of fibrinogen to fibrin

thrombocyte *n* **1** a (nucleated) blood platelet **2** a cell of an invertebrate with the function of blood clotting similar to blood platelets

thrombosis *n*, *pl* **thromboses** the formation or presence of a blood clot within a blood vessel during life

thrombus *n*, *pl* **thrombi** a blood clot formed within a blood vessel and remaining attached to its place of origin — compare EMBOLUS

throwback *n* (an individual exhibiting) reversion to an earlier genetic type or phase

throw back *vi* to revert to an earlier genetic type or phase

thrush *n* **1** a whitish intensely irritating fungal growth occurring on mucous membranes, esp in the mouth or vagina **2** a suppurative disorder of the feet in various animals, esp horses

thumb *n* the short thick digit of the human hand that is next to the forefinger and is opposable to the other fingers; *also* the corresponding digit in lower animals

thymidine *n* a nucleoside containing thymine

thymine *n* a pyrimidine base that is 1 of the 4 bases whose order in the DNA chain codes genetic information — compare ADENINE, CYTOSINE, GUANINE, URACIL

thymus *n* a gland in the lower neck region that functions in the development of the body's immune system and in humans tends to atrophy after sexual maturity

thyroglobulin *n* an iodine-containing protein that is the form in which hormones of the thyroid gland are stored

thyroid[1] *also* **thyroidal** *adj* of or being (an artery, nerve, etc associated with) **a** the thyroid gland **b** the chief cartilage of the larynx

thyroid², **thyroid gland** *n* a large endocrine gland that lies at the base of the neck and produces hormones (eg thyroxine) that increase the metabolic rate and influence growth and development

thyroid-stimulating hormone *n* a hormone secreted by the front lobe of the pituitary gland that regulates the formation and secretion of thyroid hormones

thyrotoxicosis *n* hyperthyroidism

thyrotrophin, **thyrotropin** *n* THYROID-STIMULATING HORMONE

thyroxine, **thyroxin** *n* an iodine-containing amino acid that is the major hormone produced by the thyroid gland and is used to treat conditions in which the thyroid gland produces insufficient quantities of hormones

thyrsus *n*, *pl* **thyrsi** a flower cluster (eg in the lilac and horse chestnut) with a long main axis bearing short branches which in turn bear the flowers ➔ FLOWER

tibia *n*, *pl* **tibiae** *also* **tibias** 1 the inner and usu larger of the 2 bones of the vertebrate hind limb between the knee and ankle; the shinbone — compare FIBULA ➔ ANATOMY 2 the 4th joint of the leg of an insect between the femur and tarsus ➔ INSECT

tic *n* (a) local and habitual spasmodic motion of particular muscles, esp of the face; twitching

tick *n* 1 any of a superfamily of related bloodsucking arachnids that feed on warm-blooded animals and often transmit infectious diseases 2 any of various usu wingless parasitic insects

till *vt* to work (eg land) by ploughing, sowing, and raising crops

tiller¹ *n* a sprout or stalk (from the base of a plant)

tiller² *vi, of a plant* to put forth tillers

tincture *n* a solution of a substance in alcohol for medicinal use <~ *of iodine*>

tinnitus *n* a subjective ringing or roaring sensation of noise

tissue *n* a cluster of cells, usu of a particular kind, together with their intercellular substance that form any of the structural materials of a plant or animal

tissue culture *n* the technique of growing body tissues in a culture medium outside the organism

tissue fluid *n* a fluid surrounding the cells of the tissues of the body that serves in the transport of nutrients and waste

toad *n* any of numerous tailless leaping amphibians that differ from the related frogs by living more on land and in having a shorter squatter body with a rough, dry, and warty skin

toadstool *n* a (poisonous or inedible) umbrella-shaped fungus

tobacco mosaic *n* any of several mosaic virus diseases of plants of the nightshade family, esp tobacco

tocopherol n a compound of high vitamin E potency obtained from germ oils or by synthesis

tolerance n the ability to endure or adapt physiologically to the effects of a drug, virus, radiation, etc

tolerate vt to endure or resist the action of (eg a drug) without grave or lasting injury

tomentum n, pl **tomenta** a covering of densely matted woolly hairs

tone¹ n 1 a sound of definite pitch and variation 2 the state of (an organ or part of) a living body in which the functions are healthy and performed with due vigour 3 normal tension or responsiveness to stimuli—used esp with reference to a muscle

tone² vt to impart tone to <*medicine to ~ up the system*>

tone-deaf adj relatively insensitive to differences in musical pitch

tongue n 1 a fleshy muscular movable organ of the floor of the mouth in most vertebrates that bears sensory end organs and small glands and functions esp in tasting and swallowing food and in human beings as a speech organ ☞ SENSE ORGAN 2 a part of various invertebrate animals that is analogous to the tongue of vertebrates

tonic¹ adj 1 marked by prolonged muscular contraction <~ *convulsions*> 2 increasing or restoring physical or mental tone

tonic² n sthg (eg a drug) that increases body tone

tonoplast n the membrane surrounding a vacuole in the cytoplasm of a plant cell

tonsil n 1 either of a pair of prominent oval masses of spongy lymphoid tissue that lie 1 on each side of the throat at the back of the mouth ☞ ANATOMY 2 any of various masses of lymphoid tissue that are similar to tonsils

tonsillectomy n the surgical removal of the tonsils

tonsillitis n inflammation of the tonsils

tooth n, pl **teeth** 1 any of the bony outgrowths that are borne esp on the jaws of vertebrates and are used for the mastication of food and sometimes as weapons ☞ ANATOMY 2 any of various usu hard and sharp projecting parts about the mouth of an invertebrate

toothed whale n any of a suborder of whales with numerous simple conical teeth — compare WHALE-BONE WHALE

top-dress vt to scatter fertilizer over (land) without working it in

topsoil n surface soil, usu including the organic layer in which plants form roots and which is turned over in ploughing

tori pl of TORUS

torpor n extreme sluggishness of action or function

tor

torsion *n* the twisting of a bodily organ on its own axis

torus *n, pl* **tori 1** a smooth rounded anatomical protuberance **2** a receptacle

touch¹ *vt* **1** to bring a bodily part into contact with, esp so as to perceive through the sense of feeling; feel **2** to lay hands on (sby afflicted with scrofula) with intent to heal ~ *vi* **1** to feel sthg with a body part (eg the hand or foot) **2** to lay hands on sby to cure disease (eg scrofula)

touch² *n* **1** the act or fact of touching **2** the sense of feeling, esp as exercised deliberately with the hands, feet, or lips **3** a specified sensation conveyed through the sense of touch <*the velvety ~ of a fabric*>

tourniquet *n* a bandage or other device for applying pressure to check bleeding or blood flow

toxaemia *n* an abnormal condition associated with the presence of toxic substances in the blood

toxic *adj* **1** of or caused by a poison or toxin **2** poisonous

toxicological, toxicologic *adj* of toxicology or toxins

toxicology *n* a branch of biology that deals with poisons and their effects and with medical, industrial, legal, or other problems arising from them

toxicosis *n, pl* **toxicoses** a disorder caused by the action of a poison or toxin

toxigenic *adj* producing toxin <~ *bacteria and fungi*>

toxin *n* an often extremely poisonous protein produced by a living organism (eg a bacterium), esp in the body of a host

trace¹ *n* **1** a minute and often barely detectable amount or indication **2** a vestige of some past thing

trace² *n* one or more vascular bundles supplying a leaf or twig

trace element *n* a chemical element present in minute quantities; *esp* one essential to a living organism for proper growth and development

tracer *n* a substance, esp a labelled element or atom, used to trace the course of a chemical or biological process

trachea *n, pl* **tracheae** *also* **tracheas 1** the main trunk of the system of tubes by which air passes to and from the lungs in vertebrates; the windpipe ➔ RESPIRATION **2** VESSEL **2 3** any of the small tubes carrying air in most insects and many other arthropods

tracheid *n* a tubular cell with tapering ends found in the xylem of plants that functions in water conduction and support

tracheotomy *n* the surgical operation of cutting into the trachea, esp through the skin, usu to relieve suffocation by inhaled matter

244

tract *n* a system of body parts or organs that collectively serve some often specified purpose <*the digestive* ~>

traction *n* a pulling force exerted on a skeletal structure (eg in treating a fracture) by means of a special device

train *vt* to direct the growth of (a plant), usu by bending, pruning, etc

trance *n* **1** a state of semiconsciousness or unconsciousness with reduced or absent sensitivity to external stimulation **2** a usu self-induced state of altered consciousness or ecstasy in which religious or mystical visions may be experienced

tranquill·ize, -ise, *NAm chiefly* **tranquilize** *vt* to make tranquil or calm; *esp* to relieve of mental tension or anxiety by means of drugs ~ *vi* to become tranquil

tranquill·izer, -iser, *NAm chiefly* **tranquilizer** *n* a drug (eg diazepam) used to tranquillize

transcription *n* the naturally occurring process of constructing a molecule of nucleic acid (eg messenger RNA) using a DNA molecule as a template, with resulting transfer of genetic information to the newly formed molecule — compare TRANSLATION

transferase *n* an enzyme that promotes transfer of a chemical group from one molecule to another

transference *n* the redirection of feelings and desires, esp those unconsciously retained from childhood, towards a new object (eg towards a psychoanalyst conducting therapy)

transfer RNA *n* a relatively small RNA that transfers a particular amino acid to a growing polypeptide chain at the ribosome site for protein synthesis — compare MESSENGER RNA

transform *vt* to cause (a cell) to undergo transformation

transformation *n* modification of plant or animal cell culture (eg by a cancer-producing virus) resulting in unlimited cell growth and division

transfuse *vt* to transfer (eg blood) into a vein

translate *vt* to subject (genetic information, esp messenger RNA) to translation

translation *n* the process of forming a protein molecule at a ribosome site of protein synthesis from information contained usu in messenger RNA — compare TRANSCRIPTION

translocation *n* a change of location; *esp* the conduction of soluble material from one part of a plant to another

transmit *vt* -tt- to convey (infection) abroad or to another

transmitter *n* a neurotransmitter

transpire *vt* to pass off or give passage to (a gas or liquid) through pores or interstices; *esp* to excrete (eg water vapour) through a skin or other living membrane ~ *vt* **1** to give off a vapour; *specif* to give off or exude water vapour, esp from the surfaces of leaves **2** to pass in the form of a vapour, esp from a living body

transplant¹ *vt* **1** to lift and reset (a plant) in another soil or place **2** to transfer (an organ or tissue) from one part or individual to another

transplant² *n* **1** transplanting **2** sthg transplanted

transport *vt* to transfer or convey from one place to another <*mechanisms of ~ing ions across a living membrane*>

transsexual *n* sby physically of one sex with an urge to belong to or resemble the opposite sex

transude *vi* to pass through a membrane or permeable substance ~ *vt* to permit passage of

transvestism *n* the adoption of the dress and often the behaviour of the opposite sex

trauma *n, pl* **traumata, traumas 1a** an injury (eg a wound) to living tissue caused by an outside agent **b** a disordered mental or behavioural state resulting from mental or emotional stress or shock **2** an agent, force, or mechanism that causes trauma

tree *n* **1a** a tall woody perennial plant having a single usu long and erect main stem, generally with few or no branches on its lower part **b** a shrub or herbaceous plant having the form of a tree <*rose ~s*><*a banana ~*> **2** a much-branched system of channels, esp in an animal or plant body <*the vascular ~*>

tree line *n* the upper limit of tree growth in mountains or high latitudes

trematode *n* any of a class of parasitic flatworms including the flukes

trephine *vt or n* (to operate on with, or extract by means of) a surgical instrument for cutting out circular sections, esp of bone or the cornea of the eye

treponema *n, pl* **treponemata treponemas** any of a genus of spirochaetal bacteria that grow in human beings or other warm-blooded animals and include organisms causing syphilis and yaws

trial *n* an experiment to test quality, value, or usefulness

Triassic *adj or n* (of or being) the earliest period of the Mesozoic era

tribe *n sing or pl in constr* a category in the classification of living things ranking above a genus and below a family; *also* a natural group irrespective of taxonomic rank <*the cat ~*>

triceps *n, pl* **tricepses** *also* **triceps** a muscle with 3 points of attachment; *specif* the large muscle along the back of the upper arm that acts to straighten the arm at the elbow

trichocyst *n* any of the minute hairlike stinging or lassoing organs of some protozoans

trichology *n* the study and treatment of disorders of hair growth, specif baldness

trichome *n* a filamentous outgrowth; *esp* an epidermal hair structure on a plant

trichomonad *n* any of a genus of protozoans parasitic chiefly in the reproductive and urinary tracts of many animals including human beings

trichomoniasis *n, pl* **trichomoniases** infection with or disease caused by trichomonads (eg a human vaginitis or urethritis or a bovine venereal disease)

trichopteran *n* any of an order of insects consisting of the caddis flies

trichromatic *adj* of or being the theory that human colour vision involves 3 types of retinal sensory receptors

tricuspid valve *n* the heart valve of 3 flaps that stops blood flowing back from the right ventricle to the right atrium

trifid *adj* deeply and narrowly cleft into 3 teeth, parts, or points

trifoliate *adj* having (leaves with) 3 leaflets *<a ~ leaf>* ☞ PLANT

trifurcate *adj* having 3 branches or forks

trigeminal nerve *n* either of a pair of cranial nerves that supply motor and sensory fibres mostly to the face

trilobite *n* any of numerous extinct Palaeozoic marine arthropods that had a 3-lobed body ☞ EVOLUTION

trilocular *adj* having 3 cells or cavities

trimorphic, trimorphous *adj* occurring in or having 3 distinct (crystalline) forms

trinucleotide *n* a nucleotide consisting of 3 mononucleotides in combination; a codon

triplet *n* any of 3 children or animals born at 1 birth

triploblastic *adj* having three primary germ layers

triticale *n* a cereal grass that is a hybrid between wheat and rye and has a high yield and rich protein content

trivial name *n* **1** the second part of a 2 word Latin name of an animal, plant, etc, that follows the genus name and denotes the species **2** a common or vernacular name of an organism or chemical

tRNA *n* TRANSFER RNA

trocar *also* **trochar** *n* a sharp-pointed instrument used esp to insert a fine tube into a body cavity as a drainage outlet

trochal *adj* resembling a wheel

trochanter *n* **1** a rough prominence at the upper part of the femur of many vertebrates **2** the second segment of an insect's leg counting from the body

trochlea *n, pl* **trochleas, trochleae** an anatomical structure resembling a pulley; *esp* a surface of a bone over which a tendon passes

trochophore *n* a free-swimming cilia-bearing larva, esp of marine annelid worms

troglodyte *n* an ape

trophic *adj* of nutrition or growth <~ *disorders of muscle*> <~ *level*>

trophoblast *n* a layer of ectoderm on the outside of the blastula of many placental mammals that nourishes the embryo

tropic *adj* 1 of, being, or characteristic of (a) tropism 2 of *a hormone* influencing the activity of a specified gland

tropical *also* **tropic** *adj* of, occurring in, or characteristic of the tropics

tropism *n* (an) involuntary orientation by (a part of) an organism, esp a plant, that involves turning or curving in response to a source of stimulation (eg light) — compare TAXIS 2

truncate *adj* having the end square or even <*the ~ leaves of the tulip tree*> ⟿ PLANT

trunk *n* **1a** the main stem of a tree as distinguished from branches and roots **b** the main or central part of sthg (eg an artery, nerve, or column) **2** a proboscis; *esp* the long muscular proboscis of the elephant

trypanosome *n* any of a genus of

parasitic protozoans that infest the blood of various vertebrates including human beings and some types of which cause sleeping sickness

trypanosomiasis *n* infection with or disease caused by trypanosomes

trypsin *n* (any of several enzymes similar to) an enzyme from pancreatic juice that breaks down protein in an alkaline medium

tryptophan, tryptophane *n* an amino acid that is widely distributed in proteins and is essential to animal life

tsetse, tsetse fly *n, pl* **tsetse, tsetses** any of several two-winged flies that occur in Africa south of the Sahara desert and transmit diseases, esp sleeping sickness, by bites

tube *n* **1** a hollow elongated cylinder; *esp* one to convey fluids **2** a slender channel within a plant or animal body

tube foot *n* any of the small flexible tubular parts of starfish and some other echinoderms that are used esp in locomotion and grasping

tuber *n* (a root resembling) a short fleshy usu underground stem (eg a potato) that is potentially able to produce a new plant — compare BULB, CORM

tubercle bacillus *n* the bacterium that causes tuberculosis

tuberculin *n* a sterile liquid extracted from the tubercle bacillus and used in the diagnosis of tuberculosis, esp in humans and cattle

tuberculin test *n* a test for hypersensitivity to tuberculin as an indication of past or present tubercular infection

tuberculosis *n* a serious infectious disease of human beings and other vertebrates caused by the tubercle bacillus and characterized by fever and the formation of abnormal lumps in the body

tubicolous *adj*, of an annelid worm living in a self-constructed tube-shaped case or cover

tubular *also* **tubulous** *adj* having the form of or consisting of a tube <*a ~ calyx*>

tubule *n* a small tube; *esp* a slender tubular anatomical structure

tuft *n* a small cluster of long flexible hairs, feathers, grasses, etc attached or close together at the base

tumescent *adj* somewhat swollen; *esp, of the penis or clitoris* engorged with blood in response to sexual stimulation

tumour, *NAm chiefly* **tumor** *n* an abnormal mass of tissue that arises without obvious cause from cells of existing tissue and possesses no physiological function

tundra *n* a level or undulating treeless plain with a permanently frozen subsoil that is characteristic of arctic and subarctic regions

tunic *n* an enclosing or covering membrane or tissue <*the ~ of a seed*>

tunicate¹ *adj* **1a** having or covered with an enclosing or lining membrane **b** having, arranged in, or made up of concentric layers <*a ~ bulb*> **2** of the tunicates

tunicate² *n* any of a subphylum of marine chordate animals with a simple nervous system and a thick covering layer

tunnel vision *n* a condition in which the edges of the visual field are lost, leaving good vision only straight ahead

turbellarian *adj or n* (of or being) any of a class of mostly aquatic and free-living flatworms

turgid *adj* distended, swollen; *esp* exhibiting excessive turgor

turgor *n* the normal state of firmness and tension in living (plant) cells

Turner's syndrome *n* a genetically determined condition in women that is associated with the presence of only 1 X chromosome and no Y chromosome and that is characterized by a stocky physique with incomplete and infertile sex glands

tusk *vt or n* (to dig up or gash with) a long greatly enlarged tooth of an elephant, boar, walrus, etc, that projects when the mouth is closed and serves for digging food or as a weapon

20/20 *adj, of a person's vision* normal

twig *n* a small woody shoot or branch, usu without its leaves

twilight sleep *n* a drug-induced state in which awareness and memory of pain is dulled or removed

twin *n* either of 2 offspring produced at 1 birth

twinge *vi or n* **twinging, twingeing** (to feel) a sudden sharp stab of pain

twist *vt* to wring or wrench so as to dislocate or distort <~ed *my ankle*>

twitch *n* (the recurrence of) a short spasmodic contraction or jerk

two-winged fly *n* any of a large order of insects including the housefly, mosquito, and gnat with functional front wings and greatly reduced rear wings used to control balance

tympanic bone *n* a bone enclosing part of the middle ear and supporting the tympanic membrane

tympanic membrane *n* a thin membrane separating the outer ear from the middle ear that functions in the mechanical reception of sound waves and in their transmission to the site of sensory reception; the eardrum ➔ SENSE ORGAN

tympanum *n, pl* **tympana, tympanums 1a** TYMPANIC MEMBRANE **b** MIDDLE EAR **2** a thin tense membrane covering the hearing-organ of an insect

type¹ *n* **1** a lower taxonomic category selected as reference for a higher category <*a ~ genus*> **2** a particular kind, class, or group with distinct characteristics

type² *vt* **1** to identify as belonging to a type **2** to determine the natural type of (eg a blood sample)

typhoid¹ *adj* **1** (suggestive) of typhus **2** of or being typhoid

typhoid², typhoid fever *n* a serious communicable human disease caused by a bacterium and marked esp by fever, diarrhoea, headache, and intestinal inflammation

typhus *n* a serious human disease marked by high fever, stupor alternating with delirium, intense headache, and a dark red rash, caused by a rickettsia, and transmitted esp by body lice

typical *adj* showing or according with the usual or expected traits

tyramine *n* an amine derived from tyrosine that has an action on the sympathetic nervous system similar to that of adrenalin

tyrosine *n* an amino acid that occurs in most proteins and is the parent compound from which adrenalin and melanin are formed

udder *n* a large pendulous organ consisting of 2 or more mammary glands enclosed in a common envelope and each having a single nipple

ulcer *n* a persistent open sore in skin

or mucous membrane that often discharges pus

ulcerate *vb* to (cause to) become affected (as if) with an ulcer

ulna *n* the bone of the human forearm on the little-finger side; *also* a corresponding part of the forelimb of vertebrates above fishes ➔ ANATOMY

ulotrichous *adj* having woolly or crisp hair

ultracentrifuge *n* a high-speed centrifuge able to sediment colloidal or other small particles

umbel *n* an inflorescence typical of plants of the carrot family in which the axis is very much contracted so that the flower stalks spring from the same point to form a flat or rounded flower cluster ➔ FLOWER

umbellifer *n* a plant of the carrot family

umbilical *adj* of or near the navel

umbilical cord *n* a cord arising from the navel that connects the foetus with the placenta ➔ REPRODUCTION

umbilicate, umbilicated *adj* **1** depressed like a navel **2** having an umbilicus

umbilicus *n, pl* **umbilici, umbilicuses** **1** the navel **2** any of several anatomical depressions comparable to an umbilicus; *esp* HILUM 1

umbrella *n* the bell-shaped or saucer-shaped largely gelatinous

structure that forms the chief part of the body of most jellyfishes

unattached *adj* not joined or united <~ *polyps*>

unbalanced *adj* mentally disordered or deranged

unciform *adj* hook-shaped

uncinate *adj, of a plant or animal part* having a hook-shaped tip

unconditioned *adj* not dependent on conditioning or learning

unconscious¹ *adj* **1** not possessing mind or having lost consciousness <~ *matter*> <~ *for 3 days*> **2** not marked by or resulting from conscious thought, sensation, or feeling <~ *motivation*>

unconscious² *n* the part of the mind that does not ordinarily enter a person's awareness but nevertheless influences behaviour and may be manifested in dreams or slips of the tongue

underbody *n* the lower part of an animal's body

underbrush *n, NAm* undergrowth in a wood or forest

underdeveloped *adj* not normally or adequately developed <~ *muscles*>

undergrowth *n* shrub, bushes, saplings, etc growing under larger trees in a wood or forest

underhung *adj* **1** *of a lower jaw* projecting beyond the upper jaw — compare PROGNATHOUS **2** having an underhung jaw

undernourished *adj* supplied with

less than the minimum amount of the foods essential for sound health and growth

undershot *adj* underhung

underwing[1] *n* **1** one of the posterior wings of an insect **2** any of various moths that have the hind wings banded with contrasting colours

underwing[2] *adj* placed or growing underneath the wing <*a bird's ~ coverts*>

undescended *adj, of a testis* retained within the abdomen rather than descending into the scrotum at the normal age

undulate, undulated *adj* having a wavy surface, edge, or markings <*the ~ margin of a leaf*> ➜ PLANT

unfledged *adj* not feathered; not ready for flight

unguis *n, pl* **ungues 1** a nail, claw, or hoof, esp on a digit of a vertebrate **2** a narrow pointed base of a petal

ungulate[1] *adj* **1** having hoofs **2** of or belonging to the ungulates

ungulate[2] *n* any of the group consisting of the usu herbivorous hoofed mammals

unicellular *adj* having or consisting of a single cell

unilateral *adj* produced or arranged on or directed towards 1 side <*a stem bearing ~ flowers*>

unilocular *adj* containing a single cavity <*~ anthers*>

uninucleate *adj* having a single nucleus

uniparous *adj* producing only 1 egg or offspring at a time; *also* having produced 1 offspring

unisexual *adj* of or restricted to 1 sex: **a** male or female but not both **b** dioecious <*a ~ flower*>

unit *n* determinate quantity (eg of length, time, or heat) adopted as a standard measurement ◉

unit character *n* a natural character inherited either as a whole or not at all; *esp* one dependent on the presence or absence of a single gene

unit membrane *n* a 3-layered semipermeable membrane structure consisting of a lipid layer 2 molecules thick surrounded by a protein layer on each side

univalent[1] *adj, of a chromosome* not pairing with another chromosome at meiotic cell division

univalent[2] *n* a univalent chromosome

univalve *adj* having or consisting of 1 valve

universe *n* POPULATION 3

unpaired *adj* not paired; *esp* not matched or mated

unstable *adj* not stable; not firm or fixed; not constant: eg **a** apt to move, sway, or fall; unsteady **b** characterized by inability to control the emotions **c** readily decomposing or changing in chemical composition or biological activity

uptake *n* an absorbing and incorporating, esp into a living organism

uracil *n* a base that is one of the 4 bases whose order in the polynucleotide chain of RNA codes genetic information — compare ADENINE, CYTOSINE, GUANINE, THYMINE

uraemia *n* accumulation in the blood of toxic constituents normally eliminated by the kidneys

urea *n* a nitrogen-containing compound that is present in urine and is a final product of protein decomposition

ureter *n* a duct that carries away the urine from a kidney to the bladder or cloaca

urethra *n, pl* **urethras, urethrae** the canal that in most mammals carries off the urine from the bladder and in the male serves also as a spermatic duct

urethritis *n* inflammation of the urethra

uric *adj* of or found in urine

uric acid *n* a compound that is present in small quantities in mammalian urine and is the chief excretory product of birds and most reptiles

urinary *adj* 1 relating to (or occurring in or constituting the organs concerned with the formation and discharge of) urine 2 excreted as or in urine

urinate *vi* to discharge urine

urine *n* waste material that is secreted by the kidney in vertebrates and forms a clear amber and usu slightly acid fluid in mammals but is semisolid in birds and reptiles

urinogenital *adj* genitourinary

urochord *n* the notochord of a tunicate

urodele *n* any of an order of amphibians (eg newts) that have a tail throughout life

urology *n* a branch of medicine dealing with the genitourinary tract

uropygial gland *n* a large gland that opens at the base of the tail feathers in most birds and usu secretes an oily fluid which the bird uses in preening its feathers

uropygium *n* the prominence at the rear end of a bird's body that supports the tail feathers

urostyle *n* a bony rod of fused vertebrae that forms the end part of the vertebral column of frogs and toads

ursine *adj* of or resembling a bear or the bear family

uterine *adj* 1a born of the same mother but by a different father b matrilineal 2 of or affecting the uterus

uterus *n, pl* **uteri** *also* **uteruses** 1 an organ of the female mammal for containing and usu for nourishing the young during development before birth ↗ REPRODUCTION 2 a structure in some lower animals analogous to the uterus in which eggs or young develop

utricle *n* a small pouched part of an

Length

	m	cm	in	ft	yd
1 metre	1	100	39.370	3.281	1.094
1 centimetre	0.010	1	0.393	0.033	0.011
1 inch	0.025	2.54	1	0.083	0.028
1 foot	0.305	30.48	12	1	0.333
1 yard	0.914	91.44	36	3	1

1 kilometre = 100m
1 mile = 1760yd

Capacity measure

		1 fluid ounce	=	28.4 ml
5 fl oz	=	1 gill	=	0.142 l
4 gill	=	1 pint	=	0.568 l
2 pt	=	1 quart	=	1.136 l
4 qt	=	1 gallon	=	4.546 l
		1 millilitre	=	0.002 pt
10 ml	=	1 centilitre	=	0.018 pt
10 cl	=	1 decilitre	=	0.176 pt
10 dl	=	1 litre	=	1.760 pt

Area

1 are	=	100 m²	119.6 yd²
1 hectare	=	100 are	2.471 acres
1 km²	=	100 hectares	0.387 mi²
1 acre	=	0.405 hectares	4840 yd²
1 rood	=	1011.7 m²	¼ acre
1 mi²	=	2.59 km²	640 acres

Temperature

$°Fahrenheit = (\frac{9}{5} \times \chi°C) + 32$

$°Centigrade = \frac{5}{9} \times (\chi°F - 32)$

where χ is the temperature
needing converting

Weight

		1 grain	=	64.8 mg
		1 dram	=	1.772 g
16 drams	=	1 ounce	=	28.35 g
16 oz	=	1 pound	=	0.45 kg
14 pounds	=	1 stone	=	6.35 kg
2 stones	=	1 quarter	=	12.7 kg
4 quarters	=	1 hundredweight	=	50.8 kg
20 cwt	=	1 (long) ton	=	1.016 tonnes
		1 milligram	=	0.015 grain
10 mg	=	1 centigram	=	0.154 grain
10 cg	=	1 decigram	=	1.543 grain
10 dg	=	1 gram	=	15.43 grain
				= 0.035 oz
1000 g	=	1 kilogram	=	2.205 lb
1000 kg	=	1 tonne	=	
		(metric ton)	=	0.984 (long) ton
1 slug	=	14.594 kg	=	32.174 lb

unit ⊚

Velocity

	m/sec	km/hr	mi/hr	ft/sec
1 metre per second	1	3.6	2.237	3.281
1 kilometre per hour	0.278	1	0.621	0.911
1 mile per hour	0.447	1.609	1	1.467
1 foot per second	0.305	1.097	0.682	1

Pressure

	$N/m^2(Pa)$	kg/cm^2	lb/in^2	atmos
1 N/m^2(Pa)	1	1.020×10^{-5}	1.450×10^{-4}	9.869×10^{-6}
1 kg/cm^2	980.665×10^2	1	14.223	0.968
1 lb/in^2	6.895×10^3	0.07068	1	0.068
1 atmos	1.013×10^5	1.033	14.696	1

1 pascal = 1 newton per square metre = 10 dynes per square centimetre
1 bar = 10^5 newtons per square metre = 0.987 atmosphere
1 torr = 133.322 newtons per square metre = 1/760 atmosphere
1 atmosphere = 760 mm Hg = 29.92 in Hg = 33.90 ft water (all at 0°C.)

Work and energy

	J	cal	kWhr	btu
1 joule	1	0.239	2.778×10^{-7}	9.478×10^{-4}
1 calorie	4.187	1	1.163×10^{-6}	3.968×10^{-3}
1 kilowatt hour	3.6×10^6	8.598×10^5	1	3412.14
1 British Thermal Unit	1055.06	251.997	2.930×10^{-4}	1

1 joule = 1 newton metre = 1 watt second = 10^7 ergs = 0.738 ft lb
1 electron volt = 1.602×10^{-19} joule

Force

	N	kg	dyne	poundal	lb
1 newton	1	0.102	10^5	7.233	0.225
1 kilogram force	9.807	1	9.807×10^5	70.932	2.205
1 dyne	10^{-5}	1.020×10^{-6}	1	7.233×10^{-5}	2.248×10^{-6}
1 poundal	0.138	1.410×10^{-2}	1.383×10^4	1	0.031
1 pound force	4.448	0.454	4.448×10^5	32.174	1

Base SI units

unit	symbol	concept
ampere	A	electric current
candela	cd	luminous intensity
kelvin	K	thermodynamic temperature
kilogram	kg	mass
metre	m	length
mole	mol	amount of substance
second	s	time

Derived SI units with names

coulomb	C	electric charge
farad	F	capacitance
henry	H	inductance
hertz	Hz	frequency
joule	J	work or energy
lumen	lm	luminous flux
lux	lx	illumination
newton	N	force
ohm	Ω	electric resistance
pascal	Pa	pressure
tesla	T	magnetic flux density
volt	V	electric potential (difference)
watt	W	power
weber	Wb	magnetic flux

Fundamental constants

constant	symbol	value
velocity of light in a vacuum	c	$2.998 \times 10^8 \, \text{m s}^{-1}$
charge on electron	e	$1.602 \times 10^{-19} \, \text{C}$
rest mass of an electron	m_e	$9.110 \times 10^{-31} \, \text{kg}$
rest mass of a proton	m_p	$1.673 \times 10^{-27} \, \text{kg}$
rest mass of a neutron	m_n	$1.675 \times 10^{-27} \, \text{kg}$
Avogadro's constant	L, N_A	$6.022 \times 10^{23} \, \text{mol}^{-1}$
standard atmospheric pressure		$1.013 \, \text{Pa}$
acceleration due to gravity	g	$9.807 \, \text{m s}^{-2}$
velocity of sound at sea level at $0\,^\circ\text{C}$		$331.46 \, \text{m s}^{-1}$
magnetic constant (permeability of free space)	μ_o	$4\pi \times 10^{-7} \, \text{H m}^{-1}$
electric constant (permittivity of free space)	$\epsilon_o = \mu_o{}^{-1}c^{-2}$	$8.854 \times 10^{-12} \, \text{F m}^{-1}$
Planck's constant	h	$6.626 \times 10^{-34} \, \text{J s}$
Boltzmann's constant	$k = \dfrac{R}{L}$	$1.381 \times 10^{-23} \, \text{J K}^{-1}$
universal gas constant	$R = Lk$	$8.314 \, \text{J K}^{-1} \, \text{mol}^{-1}$
Faraday constant	$F = Ne$	$9.649 \times 10^4 \, \text{C mol}^{-1}$
gravitational constant	G	$6.673 \times 10^{-11} \, \text{N m}^2 \, \text{kg}^{-2}$

Other units used with SI (in specialized fields)

unit	symbol	value	concept
ångstrom	Å	10^{-10} m	length
astronomical unit	AU	149,600,000 km	length
degree celsius	C	1 K	temperature
electron volt	eV	1.60219×10^{-19} J	energy
parsec	pc	30857×10^{12} m	length

Metric prefixes

exa	E	10^{18}	1 000 000 000 000 000 000
peta	P	10^{15}	1 000 000 000 000 000
tera	T	10^{12}	1 000 000 000 000
giga	G	10^{9}	1000 000 000
mega	M	10^{6}	1000 000
kilo	k	10^{3}	1000
hecto	h	10^{2}	100
deca	da	10^{1}	10
deci	d	10^{-1}	0.1
centi	c	10^{-2}	0.01
milli	m	10^{-3}	0.001
micro	μ	10^{-6}	0.000 001
nano	n	10^{-9}	0.000 000 001
pico	p	10^{-12}	0.000 000 000 001
femto	f	10^{-15}	0.000 000 000 000 001
atto	a	10^{-18}	0.000 000 000 000 000 001

animal or plant body; *esp* the larger chamber of the membranous labyrinth of the ear into which the semicircular canals open

uvula *n, pl* **uvulas, uvulae** the fleshy lobe hanging in the middle of the back of the soft palate ☞ SENSE ORGAN

uvular *adj* of the uvula <~ *glands*>

vaccinate *vt* **1** to inoculate with cowpox virus in order to produce immunity to smallpox **2** to administer a vaccine to, usu by injection ~ *vi to perform or practise the administration of vaccine*

vaccine¹ *adj* of cowpox or vaccination <*a ~ pustule*>

vaccine² *n* material (eg a preparation of killed or modified virus or bacteria) used in vaccinating

vaccinia *n* cowpox

vacuolate, vacuolated *adj* containing 1 or more vacuoles

vacuolation *n* the development or formation of vacuoles

vacuole *n* a small cavity or space containing air or fluid in the tissues of an organism or in the protoplasm of an individual cell

vagal *adj* of, affected or controlled by, or being the vagus nerve

vagina *n, pl* **vaginae, vaginas 1** a canal in a female mammal that leads from the uterus to the external orifice of the genital canal ☞ REPRODUCTION **2** a sheath; *esp* a leaf base that forms a sheath, usu round the main stem

vaginismus *n* a painful spasmodic contraction of the vagina

vaginitis *n* inflammation of the vagina or of a covering structure (eg a tendon sheath)

vagus *n, pl* **vagi** either of a pair of cranial nerves that supply chiefly the heart and viscera

valency, *NAm chiefly* **valence** *n* **1** the degree of combining power of an element or radical as shown by the number of atomic weights of a univalent element (eg hydrogen) with which the atomic weight of the element will combine or for which it can be substituted or with which it can be compared **2** a unit of valency <*the 4 valencies of carbon*>

valine *n* an essential amino acid that occurs in most proteins

valve *n* **1** a structure, esp in the heart or a vein, that closes temporarily to obstruct passage of material or permits movement of fluid in 1 direction only **2** any of the separate joined pieces that make up the shell of an (invertebrate) animal; *specif* either of the 2 halves of the shell of a bivalve mollusc **3** any of the segments or pieces into which a ripe seed capsule or pod separates

valvular *adj* **1** resembling or functioning as a valve; *also* opening by valves **2** of a valve, esp of the heart

vane *n* the flat expanded part of a feather ☞ BIRD

variable *adj, of a biological group or character* not true to type; aberrant

variance *n* the square of the standard deviation

variation *n* **1** divergence in characteristics of an organism or genotype from those typical or usual of its group **2** an individual or group exhibiting variation

varicose *also* **varicosed** *adj* abnormally swollen or dilated <~ *veins*>

variety *n* any of various groups of plants or animals ranking below a species

vary *vi* to exhibit biological variation

vascular *adj* of or being a channel or system of channels conducting blood, sap, etc in a plant or animal; *also* supplied with or made up of such channels, esp blood vessels <*a ~ tumour*>

vascular bundle *n* a single strand of the vascular system of a plant consisting usu of xylem . and phloem together with parenchyma cells and fibres

vascular·ize, -ise *vb* to make or become vascular

vascular plant *n* a plant having a specialized liquid conducting system that includes xylem and phloem

vascular ray *n* any of several wedges of parenchymatous tissue formed from cambium that connect xylem and phloem in a vascular plant

vas deferens *n, pl* **vasa deferentia** a duct, esp of a higher vertebrate animal, that carries sperm from the testis towards the penis

vasectomy *n* surgical cutting out of a section of the vas deferens, usu to induce permanent sterility

vasiform *adj* having the form of a hollow tube

vasoactive *adj* affecting, esp in relaxing or contracting, the blood vessels

vasoconstriction *n* narrowing of the diameter of blood vessels

vasoconstrictor *n* a sympathetic nerve fibre, drug, etc that induces or initiates vasoconstriction

vasodilation *n* widening of the blood vessels, esp as a result of nerve action

vasodilator *n* a parasympathetic nerve fibre, drug, etc that induces or initiates vasodilation

vasopressin *n* a polypeptide pituitary hormone that increases blood pressure and decreases urine flow

vasopressor *adj* causing a rise in blood pressure by constricting the blood vessels

VD *n* VENEREAL DISEASE

vector *n* an organism (eg an insect) that transmits a disease-causing agent

vegetable[1] *adj* **1** of, constituting, or growing like plants **2** consisting of plants

vegetable[2] *n* **1** [2]PLANT **2** a usu herbaceous plant (eg the cabbage,

veg

bean, or potato) grown for an edible part; *also* this part of the plant

vegetation *n* 1 plant life or total plant cover (eg of an area) 2 an abnormal outgrowth on a body part (eg a heart valve)

vegetative *adj* 1a of or functioning in nutrition and growth as contrasted with reproductive functions <*a ~ nucleus*> b of or involving propagation by nonsexual processes or methods 2 relating to, composed of, or suggesting vegetation <*~ cover*> 3 affecting, arising from, or relating to involuntary bodily functions

vehicle *n* 1 an inert medium acting as solvent or carrier for active ingredients 2 an agent of transmission; CARRIER 1

vein *n* 1 BLOOD VESSEL — not used technically 2 any of the tubular converging vessels that carry blood from the capillaries towards the heart — compare ARTERY 3a any of the vascular bundles forming the framework of a leaf b any of the thickened cuticular ribs that serve to stiffen the wings of an insect

velar *adj* of or forming a velum, esp the soft palate

veld, veldt *n* a (shrubby or thinly forested) grassland, esp in southern Africa

velum *n, pl* **vela** a curtainlike membrane or anatomical partition; *esp* SOFT PALATE

velutinous *adj* covered with fine silky hairs

vena cava *n, pl* **venae cavae** either of the 2 large veins by which, in air-breathing vertebrates, the blood is returned to the right atrium of the heart ☞ RESPIRATION

venation *n* an arrangement or system of veins in a leaf, insect wing, etc

venereal *adj* 1 resulting from or contracted during sexual intercourse <*~ infections*> 2 of or affected with venereal disease <*a high ~ rate*>

venereal disease *n* a contagious disease (eg gonorrhoea or syphilis) that is typically acquired during sexual intercourse

venereology *n* medicine dealing with venereal diseases

venom *n* poisonous matter normally secreted by snakes, scorpions, bees, etc and transmitted chiefly by biting or stinging

venomous *adj* 1 poisonous 2 able to inflict a poisoned wound

venous *adj* 1 having or consisting of veins <*a ~ system*> 2 *of blood* containing carbon dioxide rather than oxygen

vent *n* the anus, esp of the cloaca of a bird or reptile

ventilate *vt* to expose to (a current of fresh) air; oxygenate

ventral *adj* 1 abdominal 2 relating to or situated near or on the front or lower surface of an animal opposite the back — compare DORSAL 3 being or located on the lower or inner surface of a plant structure

ventricle *n* a cavity of a bodily part or organ: eg **a** a chamber of the heart which receives blood from a corresponding atrium and from which blood is pumped into the arteries ⟹ RESPIRATION **b** any of the system of communicating cavities in the brain that are continuous with the central canal of the spinal cord

venule *n* a small vein (connecting the capillary network with the larger systemic veins)

vermicular *adj* of or caused by worms

vermiculate, vermiculated *adj* 1 marked with irregular or wavy lines <*a ~ nut*> 2 full of worms; worm-eaten

vermiform *adj* resembling a worm in shape

vermiform appendix *n* a narrow short blind tube that extends from the caecum in the lower right-hand part of the abdomen

vermin *n, pl* **vermin** 1 lice, rats, or other common harmful or objectionable animals 2 birds and mammals that prey on game

vernal·ize, -ise *vt* to hasten the flowering and fruiting of (plants), esp by chilling seeds, bulbs, or seedlings

vernation *n* the arrangement of foliage leaves within the bud — compare AESTIVATION

verruca *n, pl* **verrucas** also **verruccae** 1 a wart or warty skin growth 2 a warty prominence on a plant or animal

versatile *adj* capable of moving easily forwards or backwards, or esp up and down <*~ antennae*> <*~ anther*>

vertebra *n, pl* **vertebrae, vertebras** any of the bony or cartilaginous segments composing the spinal column ⟹ ANATOMY

vertebrate[1] *adj* 1 having a spinal column 2 of the vertebrates

vertebrate[2] *n* any of a division of animals (eg mammals, birds, reptiles, amphibians, and fishes) with a segmented backbone, together with a few primitive forms in which the backbone is represented by a notochord

verticil *n* a whorl

verticillate *adj* whorled; *esp* arranged in a transverse whorl like the spokes of a wheel <*a ~ shell*>

vesicle *n* 1 a membranous usu fluid-filled pouch (eg a cyst, vacuole, or cell) in a plant or animal 2 a blister 3 a pocket of embryonic tissue that is the beginning of an organ

vespertilian *adj* of bats

vespertine *adj* active or flourishing in the evening: eg **a** *of an animal* feeding or flying in early evening **b** *of a flower* opening in the evening

vespiary *n* a nest of a social wasp

vespine *adj* of or resembling wasps, esp wasps that live in colonies

vessel *n* **1** a tube or canal (eg an artery) in which a body fluid is contained and conveyed or circulated **2** a conducting tube in a plant

vestibule *n* any of various bodily cavities, esp when serving as or resembling an entrance to some other cavity or space: eg **a** the central cavity of the bony labyrinth of the ear **b** the part of the mouth cavity outside the teeth and gums

vestige *n* **1a** a trace or visible sign left by sthg vanished or lost **b** a minute remaining amount **2** a small or imperfectly formed body part or organ that remains from one more fully developed in an earlier stage of the individual, in a past generation, or in closely related forms

viable *adj* **1** (born alive and developed enough to be) capable of living **2** capable of growing or developing <~ *seeds*> <~ *eggs*>

vibrissa *n, pl* **vibrissae** any of the stiff hairs on a mammal's face (eg round the nostrils) that are often organs of touch

villous *adj* having villi or soft long hairs <~ *leaves*>

villus *n, pl* **villi** a small slender part: eg **a** any of the many minute projections from the membrane of the small intestine that provide a large surface area for the absorption of digested food **b** any of the branching parts on the surface of the chorion of the developing embryo of most mammals that help to form the placenta

virescence *n* the state of becoming green, esp of plant organs (eg petals) that are not normally green

virgin *n* a female animal that has never copulated

virile *adj* having the nature, properties, or qualities (often thought of as typical) of a man; *specif* capable of functioning as a male in copulation

virology *n* a branch of science that deals with viruses

virulence, virulency *n* **1** malignancy, venomousness **2** the relative capacity of a pathogen to overcome body defences

virulent *adj* **1a** *of a disease* severe and developing rapidly **b** able to overcome bodily defensive mechanisms <*a ~ strain of bacterium*> **2** extremely poisonous or venomous

virus *n* (a disease caused by) any of a large group of submicroscopic often disease-causing agents that typically consist of a protein coat surrounding an RNA or DNA core and that multiply only in living cells

viscera *n pl* the internal body organs collectively

visceral *adj* of or located on or among the viscera

viscid *adj* covered with a sticky layer <~ *leaves*>

viscus *n, pl* **viscera** the heart, liver, intestines, or other internal body organ located esp in the great cavity of the trunk

vision *n* the act or power of seeing; SIGHT 1

visual *adj* of, used in, or produced by vision <~ *organs*><~ *impressions*>

visual field *n* the entire expanse of space visible at a given instant without moving the eyes

visual purple *n* a light-sensitive red or purple pigment in the retinal rods of various vertebrates; *specif* rhodopsin

vital *adj* 1 concerned with or necessary to the maintenance of life <~ *organs*> 2 concerned with, affecting, or being a manifestation of life or living beings

vital capacity *n* the breathing capacity of the lungs expressed as the maximum volume of air that can be forcibly exhaled

vitamin *n* any of various organic compounds that are essential in minute quantities to the nutrition of most animals and act esp as (precursors of) coenzymes in the regulation of metabolic processes

vitamin A *n* any of several fat-soluble vitamins found in egg yolk, milk, cod-liver oil, etc that are converted into retinal in the animal body and whose lack results in night blindness

vitamin B *n* VITAMIN B COMPLEX

vitamin B₁ *n* thiamine

vitamin B₂ *n* riboflavin

vitamin B₆ *n* (a vitamin B chemically related to) pyridoxine

vitamin B₁₂ *n* a cobalt-containing water-soluble vitamin B that occurs esp in liver, is essential for normal blood formation and nerve function, and whose lack or malabsorption results in pernicious anaemia

vitamin B complex *n* a group of water-soluble vitamins that are found in most foods and include biotin, choline, folic acid, nicotinic acid, and pantothenic acid

vitamin C *n* a water-soluble vitamin found in (citrus) fruits, spinach, cabbage, or other plant parts that is used as an antioxidant for preserving foods and whose lack results in scurvy

vitamin D *n* any of several fat-soluble vitamins chemically related to the steroids and found esp in animal products (eg fish liver oils, or milk) that are essential for normal bone and tooth structure₂

vitamin D₂ *n* a synthetic vitamin D used to treat rickets and as a rat poison

vitamin D₃ *n* the main naturally occurring vitamin D, found in most fish liver oils and formed in the skin of human beings on exposure to sunlight

vitamin E *n* any of several fat-soluble compounds found esp in leaves and oils made from seeds whose lack leads to infertility and the degeneration of muscle in many vertebrates animals; *esp* tocopherol

vitamin K *n* any of several chemically related naturally occurring or synthetic fat-soluble vitamins essential for the clotting of blood

vitellin *n* a phosphorus-containing protein in egg yolk

vitelline membrane *n* the membrane that encloses the developing embryo in an egg and that in many invertebrates acts to prevent other spermatozoa from entering

vitellus *n* YOLK 2

vitreous *adj* 1 resembling glass in colour, composition, brittleness, etc <~ *rocks*> 2 of or being the vitreous humour

vitreous humour *n* the colourless transparent jelly that fills the eyeball behind the lens ☞ SENSE ORGAN

viviparous *adj* 1 producing living young, instead of eggs, from within the body in the manner of nearly all mammals, many reptiles, and a few fishes 2 germinating while still attached to the parent plant <*the ~ seed of the mangrove*>

vivisection *n* operation or (distressful) experimentation on a living animal, usu in the course of medical or physiological research

vocal cords *n pl* either of 2 pairs of mucous membrane folds in the cavity of the larynx whose free edges vibrate to produce sound

voice *n* sound produced by humans, birds, etc by forcing air from the lungs through the larynx in mammals or syrinx in birds

voice box *n* the larynx

volant *adj* (capable of) flying

voluntary *adj* of, subject to, or regulated by the will <~ *behaviour*>

voluntary muscle *n* muscle (eg most striated muscle) under voluntary control

volva *n* a thin membrane round the base of the stem supporting the cap of a fungus

volvulus *n* twisting of the intestine upon itself, causing obstruction and pain

vomit¹ *n* a vomiting; *also* the vomited matter

vomit² *vb* to disgorge (the contents of the stomach) through the mouth

vulva *n, pl* **vulvas, vulvae** the (opening between the projecting) external parts of the female genital organs

wader *n* any of many long-legged birds (eg sandpipers and snipes) that wade in water in search of food

waist *n* 1 the (narrow) part of the body between the chest and hips 2 the greatly constricted part of the abdomen of a wasp, fly, etc

walk n **1** the gait of a 2-legged animal in which the feet are lifted alternately with 1 foot always (partially) on the ground **2** the slow 4-beat gait of a quadruped, specif a horse, in which there are always at least 2 feet on the ground

warm-blooded adj having a relatively high and constant body temperature more or less independent of the environment — compare COLD-BLOODED

warning coloration n an animal's conspicuous colouring that warns off potential enemies

wart n a horny projection on the skin, usu of the hands or feet, caused by a virus; also a protuberance, esp on a plant, resembling this

Wassermann test n a test for the presence of a specific antibody in blood serum used in the detection of syphilis

wasteland n (an area of) barren or uncultivated land <a desert ~>

waste product n material (eg faeces) discharged from, or stored in an inert form in, a living body as a by-product of metabolic processes

wasting adj undergoing or causing decay or loss of strength <~ diseases such as tuberculosis>

water n **1** the odourless colourless liquid oxide of hydrogen that is a major constituent of living matter, freezes at 0°C, boils at 100°C, is weakly ionized into hydrogen and hydroxyl ions, and is a good solvent **2** a watery fluid (eg tears, urine, or sap) formed or circulating in a living body

water bloom n an accumulation of (blue-green) algae at or near the surface of a body of water

waterborne adj supported or carried by water <~ infection>

watering place n a place where water may be obtained; esp one where animals, esp livestock, come to drink

watery adj containing, sodden with, or yielding water or a thin liquid <a ~ solution> <~ vesicles>

Watson-Crick adj of the double-helix structure of DNA <guanine involved in a ~ base pair — Nature>

wattle n a fleshy protuberance usu near or on the head or neck, esp of a bird

wave n **1** a rolling or undulatory movement or any of a series of such movements passing along a surface or through the air **2** a movement like that of an ocean wave

wax n **1** beeswax **2** any of numerous plant or animal substances that are harder, more brittle, and less greasy than fats

weak *adj* **1** deficient in physical vigour; feeble **2** lacking normal intensity or potency <~ *strain of virus*> **3** deficient in strength or flavour; dilute <~ *acid*>

weald *n* **1** a heavily wooded area **2** a wild or uncultivated usu upland region

weaner *n* a young animal recently weaned

web *n* **1** SPIDER'S WEB; *also* a similar network spun by various insects **2** a tissue or membrane; *esp* that uniting fingers or toes either at their bases (eg in human beings) or for most of their length (eg in many water birds)

webfoot *n* a foot with webbed toes

weed *n* **1** an unwanted wild plant which often overgrows or chokes out more desirable plants **2** an animal, esp a horse, unfit to breed from

weight¹ *n* **1** the amount sthg weighs **2** sthg weighing a fixed and usu specified amount **3** a numerical value assigned to an item to express its relative importance in a frequency distribution

weight² *vt* to assign a statistical weight to

wetland *n* land or areas (eg tidal flats or swamps) containing much soil moisture — usu pl with sing. meaning

wet rot *n* (decay in timber caused by) any of various fungi that attack

wood that has a high moisture content

whale *n, pl* **whales**, *esp collectively* **whale** any of an order of often enormous aquatic mammals that superficially resemble large fish, have tails modified as paddles, and are frequently hunted for oil, flesh, or whalebone

whalebone *n* a horny substance found in 2 rows of plates up to 4m (about 13ft) long attached along the upper jaw of whalebone whales and used for stiffening things

whalebone whale *n* any of a suborder of usu large whales that have whalebone instead of teeth, which they use to filter krill from large volumes of sea water — compare TOOTHED WHALE

wheat *n* (any of various grasses cultivated in most temperate areas for) a cereal grain that yields a fine white flour and is used for making bread and pasta, and in animal feeds

wheat germ *n* the embryo of the wheat kernel separated in milling and used esp as a source of vitamins

white blood cell, white cell, white corpuscle *n* any of the white or colourless blood cells that have nuclei, do not contain haemoglobin, and are primarily concerned with body defence mechanisms and repair — compare RED BLOOD CELL

whitefly *n, pl* **whiteflies**, *esp collectively* **whitefly** (an infestation of) any of numerous small insects that are injurious plant pests related to the scale insects

white matter *n* whitish nerve tissue that consists largely of myelinated nerve fibres and underlies the grey matter of the brain and spinal cord or is gathered into nerves

whooping cough *n* an infectious bacterial disease, esp of children, marked by a convulsive spasmodic cough sometimes followed by a crowing intake of breath

whorl *n* an arrangement of similar anatomical parts (eg leaves) in a circle round a point on an axis (eg a stem) ☞ PLANT

wild *adj* **1a** (of organisms) living in a natural state and not (ordinarily) tame, domesticated, or cultivated **b(1)** growing or produced without the aid and care of humans <~ *honey*> **(2)** related to or resembling a corresponding cultivated or domesticated organism <~ *strawberries*> **2** not (amenable to being) inhabited or cultivated

wilderness *n* **1** a (barren) region or area that is (essentially) uncultivated and uninhabited by human beings **2** a part of a garden or nature reserve devoted to wild growth

wild type *n* the typical form of an organism as ordinarily encountered in contrast to atypical mutant individuals

wilt¹ *vi* **1** *of a plant* to lose freshness and become flaccid; droop **2** to grow weak or faint; languish ~ *vt* to cause to wilt

wilt² *n* a disease of plants marked by wilting

wind *n* gas generated in the stomach or the intestines

windbreak *n* sthg (eg a growth of trees or a fence) that breaks the force of the wind

windfall *n* sthg, esp a fruit, blown down by the wind

wing *n* **1a** (a part of a nonflying bird or insect corresponding to) any of the movable feathered or membranous paired appendages by means of which a bird, bat, or insect flies **b** any of various body parts (eg of a flying fish or flying lemur) providing means of limited flight **2a** any of various projecting anatomical parts **b** a membranous, leaflike, or woody expansion of a plant, esp along a stem or on a seed pod

wing case *n* an elytron

wingspan *n* the distance from the tip of one of a pair of wings to that of the other

winnow *vt* to remove waste matter from (eg grain) by exposure to a current of air ~ *vi* to separate chaff from grain by exposure to a current of air

winter *vi* to pass or survive the

winter ~ *vt* to keep or feed (eg livestock) during the winter

withdrawal *n* **1a** social or emotional detachment **b** a pathological retreat from objective reality (eg in some schizophrenic states) **2** the discontinuance of use of a drug, often accompanied by unpleasant side effects

wither *vi* to become dry and shrivel (as if) from loss of bodily moisture ~ *vt* to cause to wither

withers *n pl* the ridge between the shoulder bones of a horse or other quadruped

wizen *vb* to (cause to) become dry, shrunken, and wrinkled, often as a result of aging — usu in past

womb *n* the uterus ➔ REPRODUCTION

wood *n* **1** a dense growth of trees, usu greater in extent than a copse and smaller than a forest — often pl with sing. meaning **2** a hard fibrous plant tissue that is basically xylem and makes up the greater part of the stems and branches of trees or shrubs beneath the bark

wooded *adj* covered with growing trees

woodland *n* land covered with trees, scrub, etc — often pl with sing. meaning

woody *adj* **1** overgrown with or having many woods **2a** of or containing (much) wood, wood fibres, or xylem <~ *plants*> **b** *of a plant stem* tough and fibrous

worker *n* any of the sexually underdeveloped usu sterile members of a colony of ants, bees, etc that perform most of the labour and protective duties of the colony

worm *n* **1** an annelid worm; *esp* an earthworm **2** any of numerous relatively small elongated soft-bodied invertebrate animals (eg a (destructive) caterpillar, maggot, or other insect larva)

wormcast *n* a small heap of earth excreted by an earthworm on the soil surface

wound *n* **1** an injury to the body or to a plant (eg from violence or accident) that involves tearing or breaking of a membrane (eg the skin) and usu damage to underlying tissues **2** a mental or emotional hurt or blow

wrinkle[1] *n* a small ridge, crease, or furrow formed esp in the skin due to aging or stress or on a previously smooth surface (eg by shrinkage or contraction)

wrinkle[2] *vb* **wrinkling** *vi* to become marked with or contracted into wrinkles ~ *vt* to contract into wrinkles

wrist *n* (a part of a lower animal corresponding to) the (region of the) joint between the human hand and the arm

xanthine *n* (any of various derivatives of) a yellow compound that occurs esp in animal or plant tissue

xanthophyll *n* any of several yellow

to orange pigments that are derivatives of carotenes

X chromosome *n* a sex chromosome that in humans occurs paired in each female cell and single in each male cell — compare Y CHROMOSOME

xerophilous *adj* thriving in or characteristic of a dry environment

xerophyte *n* a plant (eg a cactus) structurally adapted for life and growth with a limited water supply

xiphisternum *n, pl* **xiphisterna** the lowest segment of the sternum

xiphoid *adj* 1 sword-shaped 2 of or being the xiphisternum

xiphoid process *n* the xiphisternum

X-radiation *n* (exposure to) X rays

X-ray *vt* to examine, treat, or photograph with X rays

X ray *n* an electromagnetic ionizing radiation of very short wavelength that acts on photographic film like light

xylem *n* a complex vascular tissue of higher plants that functions chiefly in the conduction of water, gives support, and forms the woody part of many plants — compare PHLOEM

yawn *n* a deep usu involuntary intake of breath through the wide open mouth

yaws *n pl but sing or pl in constr* an infectious tropical disease caused by a spirochaetal bacterium and marked by ulcerating sores

Y chromosome *n* a sex chromosome that in humans occurs paired with an X chromosome in each male cell and does not occur in female cells

yeast *n* 1 a (commercial preparation of) yellowish surface froth or sediment that consists largely of fungal cells, occurs esp in sweet liquids in which it promotes alcoholic fermentation, and is used esp in making alcoholic drinks and as a leaven in baking 2 a minute fungus that is present and functionally active in yeast, usu has little or no mycelium, and reproduces by budding

yellow *n pl but sing in constr* any of several plant diseases caused esp by viruses and marked by yellowing of the foliage and stunting

yellow fever *n* an often fatal infectious disease of warm regions caused by a mosquito-transmitted virus and marked by fever, jaundice, and often bleeding

yield¹ *vt* to bear or bring forth as a natural product <*the tree* ~s *good fruit*> ~ *vi* to be fruitful or productive

yield² *n* the capacity of yielding produce <*high* ~ *strain of wheat*>

yolk *also* **yoke** *n* 1 the usu yellow spheroidal mass of stored food that forms the inner portion of the egg of a bird or reptile and is surrounded by the white 2 material stored in an ovum that supplies food to the developing embryo

269

yol

yolk sac *n* a membranous sac, nearly vestigial in placental mammals, attached to an embryo and containing the yolk

young[1] *adj* younger; youngest in the first or an early stage of life, growth, or development

young[2] *n pl* immature offspring, esp of an animal

zone *n* a subdivision of a biogeographic region that supports a similar fauna and flora throughout its extent

zoogeography *n* zoology dealing with the geographical distribution of animals

zooid *n* an entity that resembles but is not wholly the same as a separate individual organism; *esp* a more or less independent animal produced by fission, proliferation, or other methods that do not directly involve sex

zoology *n* (biology that deals with) animals and animal life, usu excluding human beings

zoomorphic *adj* resembling the form of (part of) an animal <*a ~ orchid*>

zoonosis *n, pl* **zoonoses** any disease (eg rabies or anthrax) communicable from lower animals to human beings

zoophilous *adj* having an attraction to or preference for animals: eg **a** adapted for pollination by animals other than insects — compare ENTOMOPHILOUS **b** *of a blood-sucking insect* preferring lower animals to human beings as a source of food

zoophyte *n* a coral, sponge, or other (branching or treelike) invertebrate animal resembling a plant

zooplankton *n* planktonic animal life — compare PHYTOPLANKTON

zoospore *n* a spore capable of independent movement

zygodactyl *adj, of a bird* having 2 toes pointing forwards and 2 backwards

zygomorphic *adj* symmetrical about only 1 longitudinal plane <*the ~ flowers of the toadflax*>

zygospore *n* a plant spore (eg in some algae), formed by union of 2 similar sexual cells, that grows to produce the phase of the plant that produces asexual spores — compare OOSPORE

zygote *n* (the developing individual produced from) a cell formed by the union of 2 gametes

zygotene *n* the stage in meiotic cell division in which homologous chromosomes pair intimately

zymase *n* an enzyme or complex of enzymes that promotes the breakdown of glucose

zymology *n* the science of fermentation

zymotic *adj* **1** of, causing, or caused by fermentation **2** relating to, being, or causing an infectious or contagious disease

zymurgy *n* chemistry that deals with fermentation processes